MAKE A
JOYFUL SOUND

By the author of

HOW DO I LOVE THEE?

VALIANT COMPANIONS

MAKE A
JOYFUL SOUND

The Romance of Mabel Hubbard
and Alexander Graham Bell

An authorized biography

by

HELEN ELMIRA WAITE

illustrated with photographs

MACRAE SMITH COMPANY: PHILADELPHIA

Library of Congress Catalog Card Number 61-14958

Manufactured in the United States of America

6111

Second Printing

The photographs in this book are published with the courtesy of the following copyright holders: the Bell family, the National Geographic Society and The American Telephone and Telegraph Company. The author and publishers are especially grateful to Gilbert H. Grosvenor for the excellent photographs of the Bells taken by himself. Dr. Grosvenor is largely responsible for the compilation of the family records upon which this book is based.

AUTHOR'S NOTE

All persons, places, incidents and conversations in this book are authentic, the source material being family letters; diaries; Dr. Bell's "Home Notes"; copies of the *Beinn Bhreagh Recorder,* the Bell Family newspaper; and other family papers filed in the Bell Room of the National Geographic Society, which were made available to me by the Bell Family.

The description of the demonstration of the first telephone in Philadelphia and of Elisha Gray's visit to Bell, given in Chapter 12, is taken from William Hubbard's letters.

While the Bell Room itself is not open to the public, exhibits covering the history of the telephone may be seen at the American Telephone and Telegraph Company, New York City; the Bell Telephone Laboratories of New Jersey, Murray Hill, N.J.; the New England Telephone Company, Boston; and the Canadian Telephone Company, Montreal.

Also, exhibits at the Alexander Graham Bell Museum of Baddeck, Nova Scotia, cover all phases of the Bell work.

ACKNOWLEDGMENTS

To the many people who made this book possible I offer my deep appreciation, with especial thanks to—

Dr. John Vedder Edwards, D.D., Assistant Pastor of the National Presbyterian Church of Washington, D. C., for his introduction to the Grosvenor family; to Mrs. Gilbert H. Grosvenor and Mrs. David Fairchild, daughters of Dr. and Mrs. Alexander Graham Bell, for their permission to write this biography and for having read the book both in manuscript and galley proofs; to Dr. Gilbert H. Grosvenor for having opened the door of the Bell Room at the National Geographic Society to me (and for having created the Bell Room itself, with its invaluable resources of Bell material and photographs). Dr. Grosvenor also read both manuscript and galleys, and he and Mrs. Grosvenor have aided me in innumerable ways.

To Dr. Graham Bell Fairchild, to Dr. Mabel H. Grosvenor, to Dr. Melville Bell Grosvenor, President-Editor of the National Geographic Society, and to Mrs. Lilian Grosvenor Jones, for sharing their memories of their grandparents with me; and especial appreciation to Mrs. Carol Grosvenor Myers and her husband, Dr. Walter K. Myers.

To the staff of the National Geographic Society, and particularly to Mrs. Vera S. Cassilly, Custodian of the Bell Room, for her untiring patience and enthusiasm. To Mrs. Jeannette Ninas Johnson, Executive Secretary of the Alexander Graham Bell Association for the Deaf, and to the staff of the Volta Bureau, both of Washington, D. C.; and to Miss Catharine Hughes.

And to the people of Baddeck, Nova Scotia, who so willingly poured out their memories to me:

To Mr. Robert Bell, cousin of Dr. Bell and nephew of Mrs. Bell; to Mr. Kenneth J. MacDermid; to Mrs. Murdoch MacDermid; to Mrs. Marian Bell MacDermid Doherty; to Mr. Angus Ferguson; Mrs. Maria Fortune; the late Hon. J. A. D. McCurdy; Mrs. Annie MacInnis; Mr. John D. MacNeil; Mrs. D. D. MacRae; Miss Garfield MacKay; Mrs. Sarah MacDonald Mackinnon; Mrs. Georgie MacLeod; Mr. and Mrs. Ralph Pinaud; Mr. Walter Pinaud, Mr. Dan Stewart.

ACKNOWLEDGMENTS

And to the Curator of the Alexander Graham Bell Museum of Baddeck, Mr. Keilor Bentley.

Also to the Alexander Graham Bell Club of Baddeck, for the loan of letters and papers written by Mrs. Bell, and for having presented an out-of-season historical meeting for my especial benefit.

And always, of course, to Miss Helen M. Vreeland!

HELEN ELMIRA WAITE

CONTENTS

CONTENTS

FOREWORD

I said to my wife, "Elsie, please give me the name of the book that has held you so spellbound all day."

"*Valiant Companions,* by Helen E. Waite," she replied. "This is an absorbing, thrilling story of the life of Helen Keller and her brilliant teacher, Anne Sullivan.

"I wish we could interest this wonderful Miss Waite in writing the love story of my mother, Mabel Hubbard, and Alexander Graham Bell. Miss Waite is one of the rare persons who understand the fearful handicap that my deaf mother conquered."

The air waves set in motion by these words were still vibrating when the postman handed Elsie Bell Grosvenor a letter. Printed on the corner of the envelope was "From Helen E. Waite."

Miss Waite wrote, "Dear Mrs. Grosvenor: My publishers have given me the assignment to write a biography of your father, Alexander Graham Bell, and I have been able to locate very little printed source material about him, but what I have makes me realize that he was very much more than merely the inventor of the telephone. I would like to make the reader know him as the *man* he was. Also I am convinced that your mother was just as important a personality as he was, and I would like to make the book her story as well as his. Would there be any old letters or papers it would be possible for me to see? Would you be willing to help me?"

Millions of words have been written and published about the genius and nobility of Alexander Graham Bell. But of Bell's beautiful and gracious wife, a most gifted woman, very little has been published. With passing years she has been fading out of the story of his life. His great fame has dazzled writers, and the brilliant light that radiates from his personality has so obscured his wife that they don't see her genius and refer to Mabel Hubbard only as "his deaf wife whom he always attended devotedly for many years."

To gifted Helen Waite the daughters of Mabel Hubbard and Alexander Graham Bell for the first time gladly made available for study and use the personal letters of the Hubbard-Bell families. All the Hubbards and Bells loved to write letters to each other. There are hundreds of letters. These we had typed and arranged years ago so that when the right person appeared she could read them comfortably.

Second only to his wife and family, Alexander Graham Bell took his greatest pride and interest in the friendships he enjoyed with hundreds of telephone men and women. He marveled at the genius of the men and women working for the giant American Telephone and Telegraph Company, and for this company's famous associates, the Bell Telephone Laboratories and the Western Electric Company. He was deeply touched by their constant devotion to him and their remembrance of him. I would like on behalf of the Hubbard-Bell family to express our warm appreciation to these companies for keeping alive the memory of Alexander Graham Bell and Mabel Hubbard.

Miss Waite has arranged that Mrs. Bell will be remembered and known for herself—a lovable, capable and charming daughter, wife, mother and grandmother and beloved mother-in-law. Her forty-eight descendants are very grateful to Miss Waite.

Most important, Miss Waite's trustworthy chronicle will place Mabel Hubbard Bell where she belongs—among the truly great, inspiring women of all time.

GILBERT H. GROSVENOR
Chairman of the
Board of Trustees,
National Geographic
Society

August 15, 1961
Beinn Bhreagh
Baddeck, Nova Scotia, Canada

MAKE A
JOYFUL SOUND

✒ 1

And Goodness and Mercy Shall Follow Me

SOMEHOW THE VERY AIR IN THE ROOM SEEMED TO HAVE GROWN colder and more oppressive after the sound of Mr. Stone's uncompromising words had died away. Gardiner Greene Hubbard had heard the same verdict so often in the past months that it echoed in his ears wherever he went, or whatever he was doing. In Boston—Philadelphia—New York— always the same! All the authorities had shaken their heads glumly, but they would add, "Perhaps if you see Mr. Stone of the Hartford Asylum . . . We have heard they have been doing experiments with articulation there."

But here were the Reverend Collins Stone and Mr. Turner, both regarding him with kindly, understanding eyes, and their verdict had been even more devastating than the others. Gardiner Hubbard thought he had been prepared to hear it, but suddenly he heard the desperate note in his own voice as he expostulated. "But Mabel can speak! We notice she is using more words each day——"

Mr. Turner closed his lips grimly, and Mr. Stone, Principal of the American Asylum, shook his head decisively. "You cannot retain her speech, Mr. Hubbard. She will be dumb in three months because she cannot hear. And if by some chance she did learn to produce words, her voice would be so unpleasant it would be painful to hear it—it would be worse than the screech of a steam locomotive! How old did you say the child is now? Five? When she is ten you can place her in a deaf-and-dumb asylum and she'll be educated by the sign language."

Gardiner Hubbard thought of his winsome little daughter just beginning to emerge from the lassitude into which she'd been plunged by the virulent attack of scarlet fever that had destroyed her hearing; how for months she had lain in such a state of complete weakness and bewilderment that the doctors had feared her brain had suffered as well as her hearing; and how wonderful it had been when at last she had begun to recognize those around her and to respond to their love and attention, and even to say a few words. Now Mr. Stone was telling him that in five more years he could consign his little girl to an "asylum"—he winced at the word—where she would be taught the language of signs!

He remembered the children he had seen here, and in the other asylums he had visited—shy, strange, unresponsive, weirdly silent, making strange and baffling gestures in the air to one another. Out of sheer desperation he asked a final question: "But—surely I recall being told that Hartford had been making experiments in teaching deaf children to speak?"

Mr. Turner was looking more dour than before, and the principal threw out his hands and gave an expressive shrug. "Yes! We did indeed! After Dr. Howe's and Mr. Horace Mann's unfortunately overenthusiastic and undoubtedly exaggerated report of what they had seen in Germany— deaf children speaking and understanding the lips of others —we made every effort. You tell him the results, Mr. Turner."

Was Mr. Turner taking an actual pleasure in the tragic verdict he was giving, Mr. Hubbard wondered, or was it simply the man's naturally dry and brittle tones that made his words ring like hail stones in his visitor's ears?

"We experimented with teaching speech here for eight years, Mr. Hubbard," the former principal was saying, "A fruitless effort. We can give the deaf a measure of speech, to be sure, but we cannot make them understand the use of vocal language. It has never been done; it never will be done. Speech isn't a part of the Hartford system now, and I hope it never will be!"

Mr. Stone was smiling benignly now. "But when your little girl is old enough to be taught, I predict she'll be very happy with the beautiful language of signs!"

Suddenly Gardiner Hubbard was swept by the feeling that he had never wanted anything so much as he wanted to get away from the American Asylum, its officers and the atmosphere of stuffy satisfaction that seemed to hover over the whole place. He reached for his hat and was on his feet in one quick movement.

"Thank you both for your patience and advice, gentlemen" —he was extending his hand to each of them in turn— "but I'm afraid I don't think much of your 'beautiful language of signs'! Speech is good enough for me!"

This certainly was not the way interviews with the Reverend Collins Stone and Mr. Turner usually ended. The officials of the famous American Asylum were not accustomed to having their counsel flatly rejected. Despite the many times they had used the word "impossible" there was a look of confidence in the eyes of the gentleman from Cambridge that hadn't been there at the start of the interview.

Mr. Turner looked dourly incredulous; Mr. Stone was frowning, and his annoyance was clear in the sudden coldness of his voice. "Your stubborn willfulness will certainly do your daughter no good, Mr. Hubbard! I urge you to resign yourself to the fact that she was destined to be deaf and dumb! I had hoped we might help you see your right course."

Mr. Hubbard gave him a fleeting smile. "I think you have, sir. I think you did it by mentioning Dr. Howe's name! Good day, gentlemen!" He bowed himself out of the room and hurried down the long corridor, but it wasn't until he had shut the great door of the Asylum behind him that he seemed able to draw a deep, free breath.

All through the late afternoon, as he watched the Con-

19

necticut hills slip past his train window, Gardiner Hubbard almost fought against the spark of new hope that Mr. Stone had so unwittingly kindled. He had followed so many false paths in the past few months that he involuntarily recoiled from the thought of another disappointment, but for Mabel...

Mr. Hubbard gripped the arm of his seat, thinking of his small daughter, bewildered and helpless in the silent world in which she was finding herself. He remembered the long, terrible weeks when she had lain frighteningly quiet, or moved on unsteady feet without seeming to notice anyone or anything about her, apparently confirming the doctors' dire beliefs that her brain had been affected. Even now Mabel's father drew a hard breath at the thought. But her young mother had refused to believe it. She had persisted in talking to the child, speaking to her eyes now instead of her ears, trying to win her attention by showing her familiar and once-loved things—and finally there had come a glorious day when she knew she had broken through and reached May's conciousness.

Shortly before her illness May had been taken to Barnum's Exhibition and had seen the famous General Tom Thumb and Mrs. Thumb. She had been so delighted with the tiny lady that her mother had purchased a photograph of the famous midget, and May would kiss it over and over, saying, "Little lady! Little lady!"

On the day that would always be vivid in the Hubbards' memory, Mrs. Hubbard had a sudden inspiration. Lifting the little photograph from the bureau, she held it before May. "Little lady, May," she said very distinctly. "Little lady!"

For the first time since her illness a slow smile dawned in the child's face. A little hand reached out to take the picture. "Lit-tle lady," the long stilled voice was repeating. "Little lady." And then, with the familiar gesture, she had kissed the pictured face. The Hubbards were exultant that day!

Under her mother's coaxing the little girl identified more

and more objects, and even volunteered a few words. It tore at their hearts when she asked, "Why don't the birdies sing? Why don't you talk to me?" But the great and wonderful thing remained: Mabel herself could speak!

Realizing that she would need special education and training, Gardiner Hubbard had promptly begun a search in schools for the deaf for a person who would be able to teach the child, help her regain and keep her speech and understand those about her. And then the bitterest blow of all descended upon the Hubbards. Mr. Hubbard was a member of the Massachusetts State Board of Education, but he was appalled and disgusted to discover that in the enlightened year of 1863 there was no teacher who could— or would—teach speech to a little deaf girl. They all seemed united in saying she never would understand the words of others. Indeed, there were no real schools; the only places for the deaf were asylums. And both Gardiner and Gertrude Hubbard recoiled at that word.

Mr. Stone and Mr. Turner, considered foremost in their field, had merely confirmed the general verdict. But Mr. Stone's remark about Horace Mann and Dr. Howe and the deaf children in Germany who could speak and understand others had set him thinking. "Unfortunately overenthusiastic and exaggerated"? Mr. Hubbard knew Horace Mann's work with the Massachusetts school system and Dr. Howe's reputation as head of the Perkins Institution for the Blind, and he couldn't imagine either the precise Horace Mann or the dynamic Dr. Howe exaggerating. If they said they had seen deaf children achieving these things it had to be true.

Horace Mann was dead, but Samuel Howe was still very much alive, and almost a neighbor of the Cambridge Hubbards, in South Boston. Had help for Mabel been so close to them while her father was combing distant cities for it?

Staring out into the gathering darkness, Mr. Hubbard told himself not to hope. Why should a director of a school for the blind hold the answer for a deaf child when experts working with the deaf had said bluntly there was no answer

21

—not the kind he wanted at any rate? And yet—tomorrow he would go to South Boston. The rumbling wheels took up the rhythm of his thoughts: "Don't hope—see Dr. Howe —don't hope—see Dr. Howe—*don't hope*——"

What Mr. Hubbard had seen of institutions made him dread visiting another, but nothing in his rounds of institutions had prepared him for Perkins, with its wide and cheerful entrance hall, the glimpses of spacious rooms, the beauty of the marble floor and stairs—so different from the drabness he usually found. What amazed him most was the alert, fearless bearing of the pupils who passed him, and the eager brightness in most of the sightless faces. He marveled when the boy who had been directed to take him to Dr. Howe led the way up the broad flight of stairs as swiftly and confidently as Mr. Hubbard was mounting himself. The boy knocked at the door at the head of the stairs, and then opened it and stood aside to let the visitor enter, without being either pitiful or groping in any of his actions, as he announced, "A gentleman to see you, sir."

"Oh, yes, Mr. Hubbard!" Dr. Howe had glanced at the card in his hand. "Come in! Thank you, Tom." He was the first director of an institution whose smile and handshake gave Mr. Hubbard a conviction that here was a source of understanding and help.

"I know of your work on the Massachusetts State Board of Education, Mr. Hubbard, and I'm glad to see you here. You said in your note on your card you wanted my advice on something urgent. If I can help——"

Gardiner Hubbard swallowed and opened his dry lips, and suddenly the story of Mabel's illness and all her parents' fears and very small hopes was rushing out. As he talked, Mr. Hubbard thought he had never realized before how hopeless the situation really was. The very telling of his quest to save Mabel from a life of silence and isolation made him realize how helpless he was. And then he had a sort of shock—for the man opposite him was not shaking his head glumly and making negative sounds; he was listening with widening eyes and a face alight with excitement.

Someone once said that when Samuel Gridley Howe met a great challenge "he kindled like a torch." Gardiner Hubbard saw him kindle now, and before the story was completely finished he was interrupting. "Of course you can save your daughter's speech—and add to it until she has a normal vocabulary! And she will not sound like a steam locomotive! I know! Oh, I know all the arguments you've heard! Especially from our friends in Hartford!" He made a grimace. "But I myself have seen children who were born deaf being taught to speak and understand conversation. I have seen it and proved it to my own satisfaction."

"Mr. Turner said something about German schools——"

Dr. Howe's voice was a little grim now. "I'll warrant he didn't remark very favorably on that report! Twenty years ago Horace Mann and I took our wives on wedding trips to Europe. While there we visited schools for both normal and afflicted children. And in Germany we saw schools where even the children born deaf used speech clearly and fluently and followed conversations from the lips of others—lip reading, it's called. And it's no fraud. I tested the pupils myself. When we came home Mr. Mann published his report on what we'd seen."

Again Dr. Howe made an expressive face. "You should have heard the outcry! And the names we were called! So I found two deaf children and taught them to speak and lip-read myself. I assure you it can be done very successfully, Mr. Hubbard, and you can accomplish it with your little girl. Mr. Mann and I had hoped to see oral schools set up for the deaf in America, but in some ways we are disgracefully backward. We like our *institutions* rather than schools."

It flashed upon Mr. Hubbard that this was the difference between the other places he'd seen and Perkins. They had been institutions—asylums. Perkins was a true school. But even with hope soaring within him, he was still questioning.

"But—if all this was published and proved twenty years

23

ago why does everyone tell me 'Impossible' now? And Mr. Turner assured me that at Hartford they had experimented eight years——"

What they did was to throw in a few half-hearted speech lessons with their precious sign language!" Dr. Howe's growl was impatient. " 'An oral department,' they said. Of course it didn't work! Mixing speech and signs never will! I was grieved to have the matter end so, but Mr. Mann had his hands too full of public school matters, and I too much to do here to press it. Opposition rose all around us——"

"Even in the face of your proof?" It seemed cruel and unbelievable, but Mr. Hubbard knew it was true.

"Yes," Dr. Howe's face saddened, "chiefly from Mr. Thomas Gallaudet of the American Asylum. He set his face against us, and the other institutions and the public followed his lead." And then the wonderful smile broke out again. "But perhaps we are beginning again with your little girl! And now let's think about the little Mabel." He leaned forward, suddenly intent and sober. "Go on just as you say your wife has been doing. Talk-talk-talk to her, just as you do to your other children. Make sure she is watching your lips. And teach her by vibration. Have her feel your throat— the cat's purr—the piano. And make her talk. Whenever she wants anything, make her speak for it. Don't let anyone use signs to her, and if she uses them, pretend not to understand. It will be heartbreaking at first, but it will pay off. And she really has one great advantage—she has heard and used her voice for the first five years of her life. In a year or so you can probably locate a teacher for her, but just now the important thing is to establish her speaking and lip-reading ability, and I think her parents can do more for her there than any teacher. Let me know how she gets on."

Mr. Hubbard knew later that he must have managed to stand up, and he hoped he had been able to murmur some kind of thanks. He knew that he had gripped the director's hand, and that Dr. Howe had nodded his understanding, his own voice suspiciously husky, as he repeated, "Let me hear!"

Somehow Mr. Hubbard had sped down the stairs; by some miracle he had boarded the right street car for Cambridge; but it was as if Dr. Howe had suddenly opened a door onto an outlook so dazzling that it obliterated everything else.

Then he was hurrying up the walk to the Brattle Street house. The door opened quickly. Gertrude Hubbard had been watching for him. She was trembling, and there were tears on her lashes, but her eyes were shining, and her voice had a joyous quality he hadn't heard in months.

"Gardiner—whatever Dr. Howe said, let me tell you something first! Gardiner dear, I know now that May understands what I say! Sometimes I hoped she did, but now I know! You remember she could repeat the Shepherd's Psalm before she was ill?" She was quivering so that her husband caught and held her fast even as he nodded. "I was repeating it to her this morning and—and when I came to the last part she—she said it with me! *'And goodness and mercy shall follow me'*—Oh, Gardiner, don't you see what that means? She remembered the Psalm, she understood my lips, and she could say it! And if she could do that, why can't we teach her other things—everything?"

$\backsim$ 2

Steppingstones for Mabel

GERTRUDE HUBBARD LOOKED WISTFULLY AT THE DOOR OF THE nursery, closed upon Mabel and her new teacher. Oh, little Berta and Grace were also there getting acquainted with her, but six-year-old Berta and four-year-old Grace always accepted everything and everyone instantly. May was standing on the threshold of a maze that might prove to be very bewildering—her education. What was happening in there? Would Miss True understand May? Would May respond to her? Mrs. Hubbard had an almost uncontrollable impulse to slip from her writing table and turn the knob softly for just one quick look. Perhaps Mary True needed help in interpreting May's imperfect speech? She half rose, and then settled back. This was an important milestone for both May and Mary True and they must pass it without the presence of May's mother, with her qualms.

She re-dipped her pen and tried to continue her letter to her cousin Evelyn, but the words trailed away. Her mind was too filled with memories of other milestones Mabel had passed in the two years since her father's memorable visit to Dr. Howe.

"Make her talk!" Dr. Howe had commanded. That had been the first joyous day the Hubbards had known since May's illlness. Gertrude Hubbard had been right! In her weeks and months of continual talking to the child and encouraging her to speak for familiar objects, she had been doing the very thing Dr. Howe promised would break through Mabel's barriers.

26

But the weeks that followed had been neither joyous nor triumphant. It hadn't been easy, for example, to hold the silver mug of milk out of reach of an outstretched little hand and watch the bewildered, troubled eyes trying to fathom why this was happening and what was expected of her, while one repeated over and over, "Say, 'I want a drink of milk,' May. 'I want a drink of milk.' Say it, dear."

Or perhaps the child would discover Thomas, the coachman, bringing the carriage to the door, and would tug at her father's coat, raising a vivid, appealing little face, and it had given him a leaden sensation not to respond to her expressive gestures, but to say, "If May wants to go in the carriage she must say 'Take me for a ride.'"

But the most difficult thing the child had had to learn was neither speech nor lip reading, but ordinary walking, for the savage attack of fever had not only robbed her of hearing but wrecked her inner ear so completely that she would never again have a normal sense of balance. She had to be taught to walk "by eye," and would always need a steadying hand in the hours of twilight or night, or on a moving vehicle.

Mabel's mother sighed. Nothing had been particularly easy for most of the Hubbard household in the first month's of May's rehabilitation. Neither of May's parents would later speak or write of that time. Gardiner Hubbard made one brief comment, "At first our little girl was very unwilling to talk"; and Mrs. Hubbard let one cry escape in a letter: "What an easy life you lead! How free from care compared to mine!"

But then, gradually, the tide had turned. Miraculously, May understood more and more of what was being said to her. She used words and even sentences more freely and voluntarily, even adding to her vocabulary words she hadn't known before her illness. She acquired the ability to walk by focusing her eyes upon some goal, so that her falls, frightening and heartbreaking in the first months, were becoming far less frequent; and she was a happy and responsive member of her family once more.

From the beginning Berta and Grace had been her constant companions, so that if a request or suggestion was given to her younger sisters, such as "Berta, bring me your coat," or "Why don't you play with your doll, Grace?" May promptly followed their example. Besides, they understood one another easily.

All the little things had been like steppingstones toward their goal. May's visit to Cousin Evelyn Salisbury in Connecticut the year before had been her first venture away from her family, and her mother had been a trifle doubtful, but her doubts had gone out of the window the day of Mabel's return.

Thomas had lifted the little girl down from the buggy, and a radiant May, carefully carrying a new doll in one hand and trying to clutch a small and very lively kitten with the other, had walked into her mother's welcoming arms.

"See new kitty Cousin gived May! See dear dolly! See her clothes!" The proud possessor of all these treasures had tried to display all their charms at once, and then added, "I just came home to get clothes washed—then I go back to see Cousin some more?"

This joyful milestone had been followed by a sharp disappointment. The Hubbards had found a teacher for May. But Miss Conklin had succeeded in making a very fleeting impression on her pupil when she vanished after a few months.

She was pleasing and dignified and self-possessed, but she had spent most of her time huddled over the register in her room.

Mary True, the new, temporary governess, was young, eager and quick. She could laugh! Both the Hubbards believed they had been guided into going to the little Maine mountain village of Bethel for this past summer of 1865. Mary True, the exceptionally well-educated daughter of the town's minister, had needed a position. She had stayed with them for a few weeks and all the children had enjoyed her, but school work would be a special challenge. Now that

May was nearly eight, her formal education had to begin. Gertrude Hubbard's throat closed suddenly on a fervent petition: "Let Mary True be the right one! Grant we have found the right one!"

When the door of the nursery opened suddenly and teacher and children tumbled out together, all four were laughing, and Mary's cheeks were flushed as she met Mrs. Hubbard's pleased eyes.

"We'd like to take a little walk," the girl told her, "and you were saying at breakfast that you wanted me to pick up some kindergarten materials. Would it be all right if I took the children, Mrs. Hubbard?"

"I would be delighted." Mrs. Hubbard nodded her approval. "I'll give you the directions where to go."

Mary felt her first surge of pride in her charges as they scampered along Brattle Street. They really were beautiful children—six-year-old Berta especially, with intensely blue eyes and flying curls; and fair Grace, wearing her angelic expression, which was somehow intensified by her white coat and bonnet. They were adorable as they tumbled over each other in the crackling crimson and gold leaves, darting ahead and then swinging back to circle around their new teacher. But when Mary looked down and met the inquiring eyes of her special charge, and felt the impulsive tightening of the little hand curled in hers, she had some qualms.

She had never seen a deaf child before—well, if the truth were told, she had never dealt with any children before—and how did one go about breaking the door of a child's silence? Especially when teaching three other lively little girls at the same time? Mrs. Hubbard had told her that morning that she was expecting a neighbor's daughter to join her own children in the classroom.

The Hubbards had told Mary frankly that May's education would be completely pioneer work, and the outcome unpredictable, since all the experts had been almost unanimous in declaring that it couldn't be accomplished as her parents wanted it done, but Mary had been ready and eager and confident. She had sensed that the Hubbards were friendly, cultured people; she liked sixteen-year-old "Sister," the oldest of the five Hubbard daughters; was enchanted with the mischievous, adorable Berta and Grace; loved May, and almost worshipped the baby of a few months, Marian.

Back in Bethel, she had been sure she could do all that was asked of her. But she was awed by aristocratic Cambridge and had been deeply impressed the night before when she stepped into the wide hall of the Hubbard home and saw "the crimson splendor of the parlor," which she later described to a friend. The walls were hung with crimson velvet paper, the beautiful mantel was of veined white marble, and the floor made an alternating pattern of light and dark diamond-shaped polished squares, gleaming in the light cast by the great glass chandelier, hung with its myriad pendants and lighted by the miracle of gas.

When a maid had ushered her into the library with its low oak bookcases and carved mantel, where the Hubbards were waiting for her, Mary's knees were behaving suspiciously, and it was just as well her hoop skirt made a good support; but Mrs. Hubbard's smile was quick, and Mr. Hubbard's welcome cordial, and "Sister" had gone with her to her room to get settled—"a lovely room with windows on three sides looking out on the garden."

Mr. Hubbard had instructed her briefly in the morning before he left for his law office. "Do all your teaching with May orally. Make her read your lips and never use a sign or gesture. And no writing."

Mrs. Hubbard had explained, "We want May and her sisters all taught together, so that she will grow up accustomed to normal people."

Mary had nodded her quick understanding of the situation. It had sounded like an excellent method; but out here with the children scampering around her, she was beginning to wonder about her own capabilities.

And then Mabel's face lit with a swift radiance, and her

queer little voice was saying, "Mit Roo, dear!" And suddenly Mary True felt her eagerness and confidence flowing back, and her answering smile sealed the bond between them which held for life. "She was my teacher for three years," Mabel remarked long afterward, "and my friend for all time."

Mabel Hubbard's first flashes of memories had been happy ones; she remembered visiting her grandfather and grandmother McCurdy's home in New York, and watching her nurse Maria making two ruffled silk dresses for her; she recalled the exciting rides in the wonderful glass-enclosed and crimson-cloth upholstered McCurdy carriage behind a "very cross Patrick"; and the thrill of the day when she was arrayed in a new red cape trimmed with what she thought was the most beautiful black and white velvet; and stepping into the hotel elevator for her first elevator ride while her father ran up the stairs to see who would get up fastest.

Perhaps the virulence of the fever had blotted out her memory as well as her hearing, for she never had any recollections of her illness or the emptiness of the months that followed. She even failed to recall being taught to speak or learning to catch words with her eyes. She always said that her connected memories began the night she sat before the fire in the nursery, rocking her doll, and saw tall young Mary True standing with her big sister in the doorway, smiling down at her, and Sister was saying, "May, Miss True has come to live with us and be your teacher."

Perhaps the child did not catch each single word, but she did get the message, and as she gravely studied her old friend from Bethel, Mary True came forward and dropped into a low chair opposite May.

"You look very happy with your dolly, May." Mabel couldn't know it, but Mary True had the gift of a very clear, distinct speaking voice. "I think I am going to like it here with you. I think we'll do a lot of things together."

Both of them confessed afterward they didn't know just

allowed me to do the best I could, and if it had not been for that, I never should have been able to teach May anything."

Sometimes she wondered how she ever achieved anything at all! Berta and Grace were irresistible, but like "little tinder-boxes, always ready to fly at each other," and Berta had an absolute genius for feats of naughtiness. It was Berta who persuaded her sisters to eat a mud pie concoction which she assured them was ice cream—Berta who took advantage of her parents' absence one morning, declared she was sick and too weak to get out of bed, and offered her worried young aunt a burningly hot little hand. Mary knew her Berta well enough to be suspicious, felt the other cool hand, and then dived under the bedclothes for the hot water bottle! Undoubtedly it was Berta who disgraced the entire Hubbard family by inveigling Grace into following their parents to the altar when baby Marian was being baptized, and playing hide-and-seek around the pulpit.

It really was hard work, dealing with normal children and May at the same time, but Mary acknowledged that Mrs. Hubbard had a point when she insisted that May share all her sisters' and Carrie Dyer's activities. If she grew up acclimated to the natual doings of normal girls, the danger of her retreating into a shell created by her prison of silence would be considerably lessened. Also being with the others, doing the same things day after day, was a painless way of increasing May's lip-reading ability and her vocabulary.

Each school session started with the Lord's Prayer and a few little songs, and shortly May was joining in the prayer and repeating the words of the songs, but it was several months before Mary realized that Mabel was actually absorbing the others' lessons by watching their lips and actions!

Excitedly she tested the child. Apparently Mabel was learning reading and spelling by observing the words and sentences on the slates as Mary prepared them, and then following the words from the others' lips. She had always been quick to learn by imitation. It was time now, Mary

33

decided, to begin special sessions of regular lessons as well as the joint ones.

May was almost galloping ahead. Now that she could read, a book was a door into an enchanted land, and her half-past-eight bedtime was an hour of doom.

"Hurry, hurry, May!" her mother or Mary would urge, "half-past eight. Off you go!"

"Just to the end of my chapter?" May would beg.

Writing and drawing came easily; May had clever little fingers and had drawn and painted busily before Mary came. Arithmetic was a bore for both teacher and pupil, but the child sat wide-eyed, breathless and enthralled over the magic the geography books were opening up.

Someone once said that Mary True could teach history to a stone post, and Mary's little pupil was far from being a stone post. Mary had been giving her a vivid introduction to the American Revolution one day when Mr. Hubbard brought a guest home for dinner, and the children were summoned to the crimson parlor to meet him.

"This is our daughter May," Mrs. Hubbard was saying, drawing Mabel forward. "May, this is Sir William Fairfax. Sir William comes from England, dear."

Mabel made the little curtsey she had been taught.

"I have read about England in my book," she volunteered.

"Have you now, my dear!" the baronet beamed delightedly, "and what did you read about England?"

"We had a war with England," Mabel informed him, "and"—her eyes shone with pride—"we beat you!"

Mr. Hubbard teased Mary for having made Mabel such a fiery patriot that she had actually insulted a guest, and she laughed, although her cheeks colored. But one morning she suddenly spoke seriously. "Actually, sir, I am sure her knowledge of American history is equal or superior to that possessed by any hearing child of her age."

Mr. Hubbard turned sober. He looked thoughtfully at May, engaged with her breakfast oatmeal, and then back at May's teacher. "You really consider that to be a fact?"

"Yes, sir, I do!"

"I confess I really would be interested in knowing how she does rank with hearing pupils," Gardiner Hubbard observed, "so possibly we can arrange to satisfy ourselves."

A few days later a Miss Ireson from the Cambridge public school system came to interview Mabel and give her the same examinations she would have taken in a regular school room.

There was no use denying that Mary True was nervous. Had she spoken too confidently? Had she been overly proud of her pupil? If she had misjudged the child's capabilities, how could she face the Hubbards' disappointment? And suppose Mabel was so disconcerted by the presence of a stranger that she couldn't show her real ability?

About that last factor she needn't have been troubled. Mabel had a natural genius for being interested in other people, and she was instantly interested in Miss Ireson and thrilled at the thought of the examination. Were these questions exactly like the ones girls in a real school would have to answer? Like the school Sister attended? Quick delight made her wiggle and then straighten in her chair with importance. She answered Miss Ireson's questions easily and delightedly.

Several days later Mr. Hubbard called Mary into the library and handed her a single sheet of paper. Mrs. Hubbard let her wool-work drop into her lap, and Mr. Hubbard's hand closed over hers. Their eyes were suspiciously bright as Mary read the report Miss Ireson had sent:

I have been exceedingly interested in examining the little Mabel, and I am happy to say she will compare very favorably with children of her own age, and is somewhat in advance of those of ten years [Mabel was nine], *who have come under my instruction. I am surprised at the readiness with which she reads from the lips, as I have never talked with her before, and she understood me without difficulty.*

35

ᴘ 3

Star Witness

FOR GARDINER HUBBARD, THE MARVEL OF EVERY JOYOUS
advance Mabel made was shot through with memories of
the situations he had found in the various asylums for the
deaf. Watching Mabel, already, at nine, well on her way
to being a natural, almost normal person, he was disquieted
by his memories of the other children of silence.

All his indignation rose over the appalling fact that Massachusetts made no provision for its silent children at all,
but dispatched them over the border to Connecticut. Worst
of all was the sickening knowledge of the empty, wasted
years a child had to spend before he was ten and was considered old enough to be accepted at an asylum at all.

"But why did Horace Mann and Dr. Howe fail?" he cried
out to his wife. "And Mr. Stone and Mr. Collins assured me
they had tried to work out teaching speech and lip reading
for eight years, and failed. Why—when we know Dr. Howe
was right?"

Gertrude Hubbard looked at him and smiled a little sadly.
"Perhaps because they didn't want to believe in it," she
suggested wisely, "and we did."

Why, that was it! Gardiner Hubbard's idea began to take
definite form then. Dr. Howe and Horace Mann had tried
to make institutions with old, long-established methods accept innovations that were directly contrary to everything
they had ever believed. The thing to do—and he intended to

36

do it—was to plan a new school, staffed by pioneer teachers unhampered by the old ideas, which would welcome children as soon as they were old enough to go to any school, where they would be given the chance to be taught as May was being taught. He would ask the Massachusetts Legislature to grant a charter for its establishment.

He had made one attempt three years before to found a school for deaf children too young to attend the Hartford school, but the effort had met defeat in the State Legislature despite the help of Dr. Howe, Frank Sanborn (Secretary of the State Board of Charities) and Mary Lamson, one of Perkins' special teachers who had seen the "German method" on her travels abroad. Messrs. Stone and Turner had come from Hartford to testify that the ideas were a "theory of visionary enthusiasts," but it had been a member of the Legislature who had dealt the bill a death blow. The Honorable Lewis Dudley stood up and informed the committee considering the bill that he had a daughter who was at Hartford, and that he was throughly satisfied with all that was being done for her there. "I am completely convinced that it is impossible to do as Mr. Hubbard says, and I would actually refuse to have my daughter taught to speak even if it were possible!"

Well, Mr. Hubbard reflected grimly, 1864 might have been a poor year to plead such a cause. The country had been at a critical point in a bitter Civil War, and the cost of soldiers and weapons would have outweighed the experiment of bringing children out of a silent world!

He'd beaten a retreat to reorganize his forces. Nevertheless, the mother of little Fanny Cushing of Boston had been inspired by the hearings to look for a teacher of the "German method" for her daughter. Mary Lamson had found her a pioneer teacher, Harriet Rogers, whose work with Fanny Cushing was so successful that Mr. Hubbard resolved to make her the head of his school. They would have to start it as a private school, but it would be a beginning. He had encouraged her to advertise for pupils, and five had been installed in the little school at Chelmsford in June, 1866. Mabel was not enrolled for two reasons: she was doing very well under Mary True, and her mother was determined that she should associate with hearing children—but the Chelmsford School had grown and its pupils were becoming natural boys and girls instead of weird, half-human little waifs. More and more influential people knew of the Chelmsford school through the efforts of Mrs. Josiah Quincy, who had invited leading members of Boston society (including several members of the Legislature) to an exhibition by Miss Rogers and her pupils at her own home.

Other friends of the new bill would be Governor and Mrs. Lippitt of Rhode Island, whose daughter Jeanie, like May, had been robbed of her hearing by scarlet fever. Taught at home by her mother, she was now a winsome girl of fifteen who could hold her own with any girl her own age. The Lippitts had exchanged encouragement and suggestions with the Hubbards and their success had been like a beacon.

It did seem as though the time was ripe to strike again. The morning Mr. Hubbard and an associate, Mr. Talbot, called on the Governor to present another petition for a school, he delighted them with the news that someone from Northampton had written to him offering fifty thousand dollars if the Commonwealth of Massachusetts would found a school for deaf children in his city. It was strange, but John Clarke had never heard of Mr. Hubbard's attempts, nor of the Chelmsford School. He only knew that he had suffered from the results of his own deafness and that he had a great concern for deaf children everywhere.

Dr. Howe and the other campaigners were elated, but after his first flash of delight passed Mr. Hubbard had a sobering thought: If Massachusetts was about to found a school for the deaf, wouldn't all the ardent advocates of the sign language be besieging the Legislature to adopt their system? Mr. Hubbard began to plan a new attack. He was a good lawyer, and he knew the value of an attack with surprise strategy.

They would have a great asset in Miss Rogers, and her reports and proofs of her success; but Mr. Hubbard's case needed a star witness, and he thought he could provide one.

On a day late in May, a day fragrant with spring perfumes and so alive with joyful bird sounds that it seemed as if it must be a good omen, Mr. Hubbard, with a flushed Mary True on one arm and a wide-eyed Mabel clinging to the other hand, walked into the special Committee room of the Boston State House.

"Good day to you, sir. . . . Mrs. Cushing, I am pleased to see you here." He was giving cordial responses to the people who turned toward him with eager greetings. "Mr. Sanborn! Yes, I do have hopes for today!"

He piloted his companions to the chairs he had selected for them, and then with his face carefully turned from May, he said quietly to Mary, "These are good seats for you. They afford a good view of the room, so May can be entertained by the proceedings; yet they're not near enough for her to pick up the words of the speakers." He smiled a trifle wryly. "She might be confused, and—well, annoyed if she discovered they do not always regard me with affection. Now as for yourself, my dear young lady, when you are called, be easy and comfortable in both your mind and your manner; tell the committee exactly how you have worked with May, and the progress you have seen her make." He smiled at her reassuringly. "Remember, there is no need for trepidation."

"N-no, sir," the girl tried valiantly to sound assured and poised, but her cheeks were hot and her unsteady voice betrayed her. After all, for a carefully-brought-up young lady of twenty-two from a simple parsonage in far-off Maine to stand before a special committee of the Massachusetts Legislature and tell them how she had done something which some of the most respected men in the country said

39

could not be done—well, it wasn't a usual or casual occurrence!

She tried to take a steadying breath, and ventured to observe, "I—I will say, sir, that you appear very confident!"

The look he gave her was half-quizzical, half-rueful as he turned to give May a final word about her behavior, and then hurried away to take his place beside Dr. Howe and Mr. Sanborn. It would be all to the good, he reflected wryly, if he were contriving to look confident, for he had seen many cases where a lawyer's air of assurance won the battle.

What was it he had told Mary True? No need for trepidation? He smiled wryly to himself. Perhaps there wasn't, but he was conscious of some unpleasant symptoms, and he had only to look toward the section where Mr. Stone and the others from Hartford were sitting, and wearing very determined expressions, to have the conviction that his battle was far from won.

Even the graduates from the American Asylum had been reading and learning of the prospect of this new miracle school, and had appeared today to add their own poignant pleas to the testimony before the committee, and as the deaf witnesses began to "talk," with their hands twisting and jerking in strange, rapid movements, translated by an interpreter, May, who sat on the edge of her chair watching the proceedings with eager eyes in an intent face, reached for her teacher's hand. "Miss True, dear, why are those people like that? Are they sick?"

Mary bent over her in swift reassurance. "No, dear, they are using their hands like that because they have never learned to talk like other people, so they talk with their hands."

"Do you know what they say?"

Mary touched the child's lips. "No, I don't," she whispered, "and you must be very quiet now."

She looked at the child curiously, wondering if May had no inkling that she herself was linked to the people she was watching with such awe. Strange, she recalled now, she had never heard anyone use the word "deaf" in May's presence.

The hearing went on and on. Both sides brought up all their arguments. Mary wished it had been proper for her to applaud Harriet Rogers when she spoke, and she had to hold her hands tightly in her lap when the gentlemen from Hartford attempted to blast her report. When she herself was summoned, Mary summoned all her courage, resolutely ignored her weak knees and hammering heart, and did as Mr. Hubbard bade her, telling her story in a low voice, simply but convincingly.

"No, I had never taught a deaf child before—I had never even seen one before I came to the Hubbards. I hadn't realized that a person not hearing wouldn't have as much language as anyone else. She didn't know what I meant in a great many cases. I have taught her in all sorts of ways, sometimes alone, but from nine to two every day with hearing children."

"What did you teach her?"

Mary True made an expressive gesture. "Arithmetic—grammar—spelling, geography and history, carrying them together—everything!"

Watching intently, Mr. Hubbard knew she had made a good impression on the committee, although they wore carefully noncommittal expressions.

The Hartford officials were under no such constraint. With every witness to the success of the Hubbard plan they grew more dour. "The recovery of speech for deaf children costs more than it's worth," Mr. Stone argued sourly. "and Mr. Hubbard's theories are right in the teeth of all experience."

"Oh, yes," the Reverend John Keep, another officer from Hartford, admitted ungraciously, "It is possible to have the deaf produce a few words, but"—he shook a bony finger at the committee and thundered a fearsome warning—"the filing of a saw and the shriek of a steam whistle combined can not produce a more disagreeable sound than is made by some of the artificial attempts at speech by the deaf.

And if all signs are excluded, as it is claimed by the small school at Chelmsford, then, however great the attainments of a deaf child may be in articulation, *his mind will still be in darkness!*" He sat down in obvious satisfaction as a murmur of dismay swept over the room.

"Well, Mr. Hubbard"—the chairman looked in his direction with lifted eyebrows—"do you care to offer a rebuttal of Mr. Keep's views?"

"I think I do, sir," Gardiner Hubbard was smiling now. "I would like to ask the committee to examine my daughter, Mabel. And I will advise you, gentlemen, to feel perfectly free to ask her any question you may desire."

After a minute of startled silence, the Chairman looked inquiringly at the other members of the committee, and catching their nods and murmurs of assent, nodded his approval. "Very good, Mr. Hubbard, you may proceed. I confess I think the child should be examined, myself."

At Mr. Hubbard's signal, Mary brought May forward. Seated in the great chair before the committee, she was an alert, vivid, appealing child, and Mr. Hubbard had a flash of pride. Never having known isolation, and belonging to a family with a large circle of relatives, May was neither embarrassed nor shy before these strange men. She had been told that they were meeting to decide whether to build a real school for her friend Miss Rogers and the deaf girls and boys in her classes, and that she could help by showing them how she could speak and understand, so she looked at them with interest.

The gentlemen themselves were suddenly feeling a little disconcerted. It was one thing to have lawyers, teachers, doctors or clergymen plead for a bill, but to have a child appear—and furthermore a deaf child . . . The various members of the committee glanced at one another and devoutly wished another man would begin.

But most of them were fathers, and after they swallowed their embarrassment, one or two began with the time-honored questions. "What is your name, child?"

Privately May must have thought that a rather stupid question; they knew who she was. But she answered politely, "Mabel Gardiner Hubbard."

Well, what did she study? Did she have brothers and sisters? And May answered them. Her voice was not normal or perfect—it had a tendency to be a little high and blurred—but it was intelligible, and it certainly was not a combination of a saw and a steam whistle! After their first astonishment the men took Mr. Hubbard's suggestion and plied her with questions in history and geography, and gave her simple problems in arithmetic. May's answers were prompt, while her whole face lit with eagerness.

"I wouldn't venture to say her mind is in complete darkness," muttered one member, while another was asking, "Can you read, my dear?"

Mary had provided a book in case May grew weary and restless during the hearing, and she brought it forward. Opening it, May read a page or two easily and clearly. Something like awe seemed to drop over the room. Most of the committee had sudden difficulty with their spectacles, but the Honorable Lewis Dudley rose abruptly and turned away. On an impulse Mr. Hubbard followed him. Meeting his understanding eyes, Mr. Dudley choked. "Do you think I could ever hear *my* daughter say just one word? Do you suppose I could ever hear her say 'Father'?"

Mr. Hubbard fumbled for his handkerchief and blew his nose hard. "I know exactly how you feel," he assured him. "Take her to Miss Rogers and see!"

The special committee requested the Legislature to pass two bills: One granted a charter for the establishment of the Clarke School; the other provided for the teaching of deaf children from five to ten years old at Clarke or other schools with funds to be supplied by Massachusetts, and in all schools the children were to be taught to speak and lip read.

Still some of the legislators held back. The matter seemed to be swaying back and forth until Lewis Dudley stood up and made a broken little speech.

43

"You know, gentlemen, how I spoke to you three years ago—how I told you of my daughter Theresa, one of the advanced pupils at the American Asylum, and how I would refuse my consent to have her taught speech if the possibility arose. I have witnessed a miracle. Mr. Hubbard's nine-year-old daughter Mabel is far more advanced and superior in every respect to my thirteen-year-old Theresa. His daughter speaks! She understands the speech of others. She is a part of the world around her. And mine"—his voice faltered and saddened—"mine is isolated—locked out of everything. Gentlemen, I tell you, for the future of all children like our daughters, give Mr. Hubbard the charter and the funds for his school!"

The legislators looked at one another silently, but when the vote began, one after another thought of the two little girls, the one who was triumphing over her silence, and the other who was still held fast, and when they lifted up their voices to vote it was as though they were voting that Theresa Dudley, like May Hubbard, should become a maker of joyful sounds.

4

Professor of Visible Speech

THE FIRST OF JUNE, 1867, PROBABLY WAS A DATE MR. HUBBARD never forgot. The establishment of the Clarke School was assured, and more, May's appearance before the Committee had won Mr. Dudley as an ardent friend and advocate for the oral cause.

Even after seeing and hearing Mabel, Mr. Dudley was more than doubtful about the possibility of his own daughter's achieving speech. He had proof now that some deaf children did accomplish it, but Jeanie Lippitt and May Hubbard and a few of Miss Rogers' pupils had been hearing and talking before they were deafened; Theresa had been born to silence. That probably made all the difference in the world.

But, at the Hubbards' urging, Mrs. Dudley did take Theresa to the little school at Chelmsford, and when Mr. Dudley visited them he was welcomed with the greeting he had never expected to hear. Theresa was calling him "Father."

"I had no more idea she could learn to talk than that I should receive the gift of tongues!" he declared, as she proudly said her precious word and the two or three others she had mastered, over and over. Henceforth his enthusiasm for the oral method would be equal to the Hubbards'.

One question asked by someone on the committee had perplexed and troubled May: "Are you a deaf child?"

She had hesitated, and then looked questioningly at her father, and in obedience to his nod had faltered "Yes." But Mary True had noticed that she was unusually silent on the ride home from the State House, and was wearing the expression that made her father nickname her "Baked Apple." Mary was not too surprised when the question came. "Miss True, dear, what is a deaf child? What did the man mean?"

Looking down into the troubled little face with its questioning eyes searching hers, Mary realized that her earlier surmise had been correct: Incredible as it seemed, May did not realize she was "different." Perhaps this had stemmed from the incessant effort her parents had made to keep her in constant association with hearing people and joining in all her sisters' activities. This had given May a happy unconsciousness of her limitations, but Mary had a quick pang as she wondered how the realization would affect the child now.

She tried to explain as simply and as gently as she could, adding that Jeanie Lippitt, whom May greatly admired, and also Fanny Cushing, with whom she had often talked, were also deaf girls, as were all the children in Miss Rogers' school. She explained that that was the reason May's father was so interested in the new school.

There were several other experiences that helped take May out of her carefree childish days and set her on the path to growing up. She was already attending dancing school. Despite her double handicap, May was a good dancer, absorbing the rhythm of the music through her feet on the floor. And she would always possess one advantage over everyone else: because of her ruined middle ear she could not be made dizzy!

The school examination May had passed so successfully proved that she was ready and able to undertake a normal and even advanced school program. There was no question of any other teacher's taking over Mary True's place now, but in a couple of years Mabel would be attending private school with her sisters, and then Mary's work with her would probably be over.

The experience that had made the deepest, most lasting impression had come last summer, when Mary and Mabel had returned to Bethel for a vacation with the True family. The Hubbard nurse, Ria, had brought baby Marian in her carriage down to the station to see them off. Mary had lost her heart to the little thing long before—she was an enchanting small creature with dark curls and brown eyes and laughing ways, and it was plain May adored her. Indeed, she flung her arms about her tiny sister, breaking into tears and declaring she had changed her mind and didn't want to go to Bethel and leave her.

Ria laughed heartily. "She'll be here when you come home, honey," she assured the tearstained May, and Mary comforted her with the reminder, "it will only be a fortnight, you know, dear!"

Bethel was an enchanted spot for May Hubbard that summer. Dr. True, Mary's father, had a great collection of queer-shaped stones lining the corridor linking the old Maine farmhouse with the barn, and they fascinated the child. So did the flock of chickens on the farm. She was given a large white one for a pet, and promptly named it Violet. One day Violet flew to the top of a post just as May appeared with his grain, and lifted his head in a long, proud crow. "Why," she cried in astonishment, "my Violet is a boy! Well, I'll call her Robbie."

Searching for eggs one day, she came upon a setting hen, gave it what the hen considered too much personal attention, and was pecked on the nose for her pains. "Retribution filled her wrathy soul," an amused Mary True wrote. "She came into the house, melted some wax, and armed with the hot wax and a cork and some cold potato, went out to seal up the hen's bill! Of course it wasn't feasible!"

Mary's brother Alfred often teased May unmercifully, and her only defence was to cry out, "You are a very, very, very worthless man!"

But before the fortnight was over the thing happened that robbed the summer of all its joyousness. Years later Mabel said she still trembled when she thought of it. A letter addressed in Mrs. Hubbard's handwriting came for Mabel, who was enchanted with letters. She opened this one carefully, but the first words frightened her: "I have something to tell you which you will be very sorry to hear . . ."

Chill foreboding seized her. She would not read any more, but somehow she knew what her mother said. She found Mary in the kitchen, and silently followed her about as she did the morning work, and then when her teacher was finished, stood with quivering lips while Mary True read the letter.

"I think we will go and look at the oak tree," Mary told her when she finished. She held out her hand and caught May's outstretched one tightly. Once settled in the favorite place in front of the house, she gathered the child into her lap and began to speak slowly.

"May, look at the blue, blue sky above the branches. Heaven is above that, and it is very, very lovely. No one is ever sick, or tired or afraid there. May, Marian has been very sick indeed, for just a few hours, but she is wonderful now, because, you see, God wanted her in Heaven, and He took her there."

For a moment Mabel's one thought was of the baby's happiness in the beautiful place Mary mentioned, and then the realization of what it meant sent a terrible wave of sorrow over her, and clutching Mary, she sobbed with a great, choking, soundless despair. Mary's arms were about her and Mary quivered too. Later she called it "one of the great griefs of my life." She had adored little Marian, and had personally taught the baby to walk. Perhaps Marian's going took a part of Mabel's childhood that nothing could replace.

There were happier incidents, too, in the process of growing up. One day Mabel armed herself with a great bouquet

*The three Bell boys, from a watercolor by their
mother. Alexander is the child with crossbow.*

*Mabel Gardiner Hubbard at 4¾, just beginning
to adjust to her baffling world of silence.*

Mabel Gardiner Hubbard as a schoolgirl in Germany, aged 14.

Alexander Graham Bell at the time he invented the telephone and became engaged to Mabel Hubbard. He was 29.

Boston Day School for the Deaf, June 21, 1871. Upper right: Alexander Graham Bell. Fourth row: 2nd from left, Sarah Fuller; 4th, Mary True.

Mabel Gardiner Hubbard, aged about 8.

Young Aleck, left, with his father and Grandfather Bell.

of pink roses to offer her friend Edith Longfellow's father when she went to ask him to sign her autograph album—and to the envy of her friends, she came back with an entire poem written on the page!

Suddenly Mabel Hubbard was twelve years old, and her father decided the time had come for her to enter Miss Sawyer's School with her sisters. By this time Mary knew that her life would be given to deaf children, and she had accepted a position with the Horace Mann School for the deaf in Boston. It gave her a curious empty feeling to think of leaving the Hubbard family, and she gasped with amazed delight when Mrs. Hubbard said gently, "If you can be happy here, I don't like to have you go somewhere else to live. And I should be glad if you could help May with her lessons after school."

May herself was fascinated with school, and a little proud to be the youngest in all her classes. There was only one subject she was unprepared for—Latin. Miss True had not taught that. So she had private lessons in German, instead, and found them easy. After a few months, however, it became evident that May and the school were not as well suited to one another as the Hubbards had hoped. Mary True had taught Mabel and two or three others together, but she had always made it a point to face Mabel and have her near enough so she could lip-read easily. At Miss Sawyer's this wasn't always feasible.

Also, a plan had been crystallizing in Gertrude Hubbard's mind. Mabel's lip reading was excellent, but her speech, while fluent and understandable, might be clearer; her voice might be sweeter and more flexible. Why not take the child to one of the special oral schools in Germany and see what could be accomplished? The Hubbards left Cambridge in March, 1870, for a visit to Washington. Anxious for May to see everything to make up for what she couldn't hear, her mother managed to take her to a White House reception,

49

where they saw President Grant, who "looked exactly like his picture, and as still!" Mabel commented.

In May they were on their way to Germany, where they toured the schools and Gertrude Hubbard discovered how much she and Mary True—both of them untrained, and using only their own imagination and wit—had accomplished. The directors of the schools they visited were incredulous when they observed May. One director refused to believe she was deaf.

"It is impossible, madam, impossible! No deaf child could possess the knowledge she does, or talk so freely!"

They tested her in every possible way, sending her, for example, up a flight of stairs, and speaking sharply while her back was turned. Finally the school head shrugged.

"I am convinced, madam, yes! But we can do nothing for her here. I tell you no German child in any oral school can match her in any way—speech, or speech-reading, or everyday knowledge. It is a true miracle."

So Mabel never attended a school for the deaf in Germany, but her mother did locate an excellent day-school for normal children that was willing to accept her. It must have been quite an experience, for no one spoke any English, not even the teachers. Apparently May had a natural flair for languages, for in a few months she was reading, writing and speaking German with ease. She said later that she "even thought in German." Neither of her parents knew the language, so when they went shopping or sight-seeing May proudly acted as translator.

She must have had a full schedule, for she wrote to a cousin, "I have philosophy, history, natural history, geography, German grammer, drawing and writing. It does not seem much, does it, but it takes up all my time from 8:30 a.m. until 5 or 6, with only a two hour recess in the middle of the day for dinner and an hour's play. I am writing now in the recess, and it will soon be time for school again."

Besides all this, May had private lessons in speech from a

50

teacher from a school for the deaf in Hanover, and drawing lessons as well. And she was learning something more—how to adjust to people outside the circle of her own family and friends. In Cambridge it would have been very easy and pleasant for her to be with people she knew, but perhaps Mrs. Hubbard had been farsighted enough to see that being thrown into the company of complete strangers would be a valuable part of her growing up, even if it wasn't always pleasant. When she was fifteen, Mabel confessed to her diary, "As I grow older I feel my loss much more severely. At home I don't remember the idea entering my head to wish to hear. But now that I am thrown into the society of strangers I am somewhat discontented." And then the high heart she would always possess asserted itself. "Only somewhat, thank God, I am getting to understand more strangers without help."

Besides attending school there were wonderful trips to take, exciting things to do and glorious places to see. The Hubbards stayed abroad two and a half years, and it was while they were in Vienna that they heard an interesting bit of news. Harriet Rogers had begged a year's leave of absence from the Clarke School to come to Germany and Austria and study the oral schools for the deaf there; and she poured out to Gertrude Hubbard and Mabel the stories she had been receiving from America concerning a young man who had lectured and taught at both the Horace Mann School and the Clarke School—a Professor A. Graham Bell, who demonstrated a new system to teach the deaf to speak and improve their voices. "Visible Speech," he called it, and from all the reports Miss Rogers had had from both Sarah Fuller, the principal of Horace Mann, and the teachers she had left in charge at Clarke, the results were real miracles. Privately May Hubbard thought that Professor A. Graham Bell sounded like a quack doctor.

Whether Miss Rogers' reports were responsible Mrs. Hubbard never said, but within a few weeks the Hubbards were back in Cambridge, and after that things moved rapidly.

51

One afternoon May found herself walking down Beacon Hill, in Boston, toward the rather dingy Boston University, piloted by her dear Miss True, on her way to the professor's class!

Amazingly, Miss True's enthusiasm was even more intense than Miss Rogers' and Miss True had actually worked with the man.

"May, all the world—the world that stammers or is deaf— is talking about this man! He's a Scot who came to Boston from Canada a year or two ago with this 'Visible Speech' his father invented. He draws the symbols on the blackboard, and by following them anyone can produce any sound in the world! Oh, don't look skeptical. I was, but he convinced me. And I have seen the two little deaf boys whom he taught to speak. And he has the marvelous faculty of making even children comprehend what he wants of them!"

Because it was Miss True who was brimming over with such enthusiasm, and she could sense that her teacher was completely convinced of this odd professor's powers, May nodded respectfully, although inwardly she still thought he sounded like a freak.

"Well," she conceded, "perhaps I will take a few lessons of him. If my voice could be a little better it will help me someday when I'm a nice lawyer's wife."

They had turned up the walk leading to the university now, and Miss True led the way quickly to a rather nondescript, dark-green room. The windows looked out onto the cemetery and the gravestones of Benjamin Franklin's father and mother. May shivered. Well, this wasn't the most cheerful place she had ever been in!

And then a tall figure came from the dark shadows of the room, and Miss True was saying, "Professor Bell, I have brought my former pupil, Miss Mabel Hubbard. Mabel, this is Professor Bell."

Professor Bell's quick smile and handclasp were the friendliest May had ever experienced. And when he spoke she realized that here was a person wonderfully easy to understand.

"I have been hearing a great deal about you from Miss True, and I have long wanted to meet you, Miss Mabel!"

Mabel gave him a swift, appraising scrutiny. He was tall and painfully thin, and looked about forty, with hazel eyes —she preferred blue—and a trick of brushing his fingers over the thick lock of hair that curled like the crest of a wave high on his forehead. Used to the dapper and polished Harvard young men who called on her two young aunts and Sister, and the impeccable gentlemen she had seen abroad, Mabel was silently scornful of Professor Bell's clothes. They were absolutely frightful, for one thing, and they didn't fit.

Professor Bell was making a quick gesture to Miss True, and she led the way down a flight of steep, gloomy stairs to a basement classroom. May gathered her skirts around her with a little shudder of distaste. Mabel Gardiner Hubbard wasn't used to going to a class in a basement.

When the little group had assembled, Professor Bell stepped to the blackboard, sketched the profile of a man with quick, assured strokes, and proceeded to add lines illustrating the vocal cords and tongue. His lithe motions and graphic demonstrations of what he wanted the little group to understand were fascinating, and he was so full of eagerness and vitality that even the skeptical fifteen-year-old girl in his audience yielded to his magnetism.

"I was forced to like to listen to him," she confessed in her diary; but then, with the stubbornness which was sometimes characteristic of her, she added, "himself I *did not like!* He dresses carelessly in a horrible, shiny broadcloth, which makes his jet black hair look shiny, too. Altogether, I do not think him exactly a gentleman. I make a distinction between *teachers* and gentlemen of means, lawyers, merchants, etc."

Did some strange, faint premonition touch her suddenly? For she was inspired to add, "*I* could never marry such a man!"

5

Boy with a Speaking Machine

DESPITE HER SWEEPING DISAPPROVAL OF HIM, MAY FOUND herself going to the little room in the Boston University to take daily lessons of Professor Bell. Her father met him, and came away liking and respecting the man, and with Mary True's enthusiasm, that satisfied Mrs. Hubbard. Perhaps here was the means to give Mabel's voice a clearer, more natural quality. Once she was well started with Professor Bell and his system, May acknowledged to her sisters that the lessons were interesting and—agreeable. She even began talking glibly about the mysteries of Visible Speech itself.

This Visible Speech was a system of a series of cryptic-looking characters which could be combined or arranged to represent any known sound in the world. It looked like some undecipherable cuneiform writing found in an ancient tomb, but it really had been invented by Professor Bell's father, Alexander Melville Bell, who was a "corrector of defective utterance," to show his stuttering pupils the positions of the palate, tongue and teeth he wished them to take to improve their speech, and his son had brought it to Boston to help him with his deaf pupils. Weird though May thought it was, she found that she could learn to decipher its symbols and produce the required sounds. And she had to confess she was enjoying the process. Mr. A. Graham Bell was one of the rare people who can make learning anything at all seem thrilling.

54

And then, just when Mabel was beginning to look forward to her daily sessions, Professor Bell suddenly informed Mrs. Hubbard that he would no longer teach her daughter himself, but would instruct his assistant, Miss Abbie Locke, how to continue her lessons. All Mrs. Hubbard's bewildered and distressed questions and pleadings failed to extract a satisfactory explanation from him.

"Oh, yes, indeed I do find her progress most promising," he assured her. "No, I have no fault whatever to find in her, and I will gladly supervise and be present at her lessons whenever possible, but it seems unwise for me to teach her personally at this time."

The truth was (although he wasn't able to confess it for two years) that he was undergoing a very disquieting experience. He had fallen headlong in love with this fifteen-and-a-half-year-old pupil.

Whichever way he seemed to turn, a new objection or obstacle reared its sinister head. Mabel Gardiner Hubbard was the daughter of one of the wealthiest men in New England, prominent and influential, so he had been told, in Washington circles as well as Boston; and while the Bells had been highly respected both in Edinburgh and London, Alexander Graham Bell was a very poor teacher in an obscure school, and his grandfather had originally been a shoemaker. Mabel Hubbard was bonnie and blithe and outgoing; and except when he was stirred by his teaching or his music, he was shy and retiring.

Also, Aleck had been under the shadow of a grave illness which had left him looking a dozen years older than he actually was. Despite May's unflattering guess that he was around forty, he was only twenty-six; but looking older would hardly endear him to a young girl, especially when she was so extremely young herself! Well, that was one obstacle time would remove—she would grow older. But there was another, and Graham Bell acknowledged it was looming the greatest of all: the girl was deaf, and he was passionately fond of music, a trained musician, and intensely interested in everything to do with sound, and for this particular girl the gates of sound were forever shut and locked. Moreover he knew what it would be like to have a deaf wife, for his own mother's hearing had been ruined by an infection.

He was a distraught young man while all the difficulties rose and battered against his soul, but the day he won through and made what he knew was his inevitable decision, he knew the first step would be to cease teaching Mabel Hubbard. "Because," he promised himself, and the smile that was one of his greatest assets must have lighted his hazel eyes even though no one was there to see it, "someday I want her to think of me as a lover—not a teacher!"

He thought how far the course of his life had swerved from his original goal of becoming a concert pianist, back in the days of his Edinburgh boyhood; and how through illness and tragedy his destiny had led him to open a little school in an American city, and to meet this particular American girl.

Nothing about Aleck Bell's boyhood had been either ordinary or tame. The Bell flat on South Charlotte Street in Edinburgh rang with the shouts of the three Bell brothers, Melville, Aleck (he would always be Aleck to his family) and Edward; and they had grown up in a way that must have been startling to their neighbors. Other boys of the Bell family's social position were dispatched to proper boarding schools to be taught to follow time-honored and orderly traditions; but the three Bells lived at home, attended local private schools or were taught by their mother, and roamed the countryside. It was an extraordinary way of bringing up boys, but it had to be acknowledged that their parents were rather extraordinary people.

Eliza Grace Bell's portraits and miniatures were highly prized. Alexander Melville Bell was a famous elocutionist, who gave exceptionally good public dramatic readings from Shakespeare and from that new author, Charles Dickens (he once got himself into serious trouble with his church for reading "from the flippant works of Mr. Dickens"), and he had amazing success in treating stutterers and others with speech troubles.

Alexander Bell took his small sons into his confidence as he worked out his system of writing sounds by drawing symbols to show the shape of the lips and tongue for different sounds, explaining how it would help people who stuttered to know just how to manage their tongues, teeth and lips. Few fathers of that era would have asked small boys to be their assistants in genuine scientific experiments.

"I would like to have you assist me at my lecture tonight," he would say. "There will be interesting people in the audience, and I think you will enjoy it. You will wait in the anteroom, out of all chance of hearing, while I explain about Visible Speech and ask for sounds to be written in it. Then you will be brought in and read the symbols and utter the sounds."

It was no wonder Melville and Aleck were thrilled. They adored their father, had learned Visible Speech readily, and were so enthralled by the challenge of translating it that they never thought of being nervous or embarrassed before an audience. And the audiences delighted in emitting the most uncanny and peculiar noises imaginable. Sometimes, even as they maneuvered their tongues and palates into the right positions, the boys themselves were utterly baffled.

Once Aleck made a loud, rasping noise that made no sense to him, but the audience cheered. Someone had suggested the noise of sawing wood.

Then his father pointed to another symbol, the most peculiar he had ever made. It called for coiling his tongue backward so the tip touched his soft palate, and then saying "T." It wasn't easy, but a gentleman sprang to his feet, applauding. "Ladies and gentlemen, I am a professor employed by the Indian Civil Service to teach young men Sanskrit. The sound I gave Mr. Bell is the Sanskrit cerebral T which I have been very unsuccessful in getting my students to pronounce, and yet Mr. Bell's son has given it correctly without ever having heard it at all!"

All of this Aleck would report to his mother with joyous pride in his father's achievements. He was the one person who did not use her long black ear trumpet, but spoke in a normal tone, with his lips against her forehead.

"And Papa says such incidents have given him a new idea," he finished triumphantly. "He thinks that if Melly and I can read and utter sounds we've never heard, why couldn't Visible Speech help deaf people learn to talk?"

Despite the fact that they knew the mysteries of their father's Visible Speech and knew more about stuttering and kindred speech ills than most doctors of their day, the Bells were very real boys with ample freedom for their own affairs. Aleck had a collection that was the pride and joy of his heart. Birds' eggs or butterflies would have been much too tame for him; he had a fine assortment of small-animal skeletons—toads, mice, rabbits, even cats and dogs—all nicely arranged and classified. It gave his mother the horrors, especially after he acquired its crowning glory, a gift from his father—a human skull.

Even in his name he wanted to be a staunch individualist. Melville was Melville James, and Edward was Edward Charles, but Aleck had been named simply "Alexander" for his grandfather, and he wanted some name that would distinguish him from the other two Alexanders. When the hospitable Bells had a guest, a friend of Mr. Bell's, who was introduced to the boys as "Mr. Alexander Graham, from Cuba," Aleck went about murmuring the name to himself, sampling it on his tongue. The musical rhythm of it pleased his sensitive ear. "Graham . . . I like that. Alexander Graham —Bell!"

On his eleventh birthday, a little later, Aleck announced to his family that he was adopting Mr. Graham's name, and that from now on he would be Alexander Graham Bell. If his parents were surprised or amused they made no objections. They were teaching their boys to think and decide for themselves, and they never ridiculed them.

Famous scientists came to discuss sound and speech with the boys' father, and sometimes stopped to talk with them, and Aleck's thoughts naturally started him on the road to scientific experiments. One day his brothers discovered him kneeling beside the family terrier, forcing it to open its jaws several times.

"Is he sick?" little Edward demanded in alarm.

Aleck shook back his black curl. "Oh, no, he's not sick. But I happened to wonder why we couldn't teach him to talk? And I think we could. Listen! When I press his jaws here—and here—he says 'ga-ga-ga.'"

Melville's quick ear caught the sound too, and at once he was down beside his brother, peering and pressing. After days of practice to find exactly the right way of manipulating the dog's jaws, they succeeded in coaxing him to produce something that sounded (if one used an ounce or two of imagination) like "How—are—oo—grandma-ma?" The fame of the Bells' talking dog spread far and wide—and oddly enough, Aleck testified later, the dog actually enjoyed the process.

But the talking dog was far from being the brothers' greatest feat. Some time later Sir Charles Wheatstone, one of Mr. Bell's scientific friends, sent him an invitation to come and see an apparatus he had built. Sir Charles had stumbled upon an old book by a German baron who claimed he had built a "Speaking Machine." Sir Charles had followed the diagrams in the book, and wanted Melville Bell to see and hear the results. He told him he might bring his son Aleck.

The apparatus did speak, though not very successfully. Aleck was disappointed about that, but he sat breathless and intent during the demonstration, and an idea was dawning in his mind. At the end of the session Sir Charles turned to him kindly. "Interested, eh boy?"

At the quick light in Aleck's face and his hurried "Oh—yes, sir!" he smiled.

"Yes, after what you did with your dog, I rather believe you would be! Suppose I lend Father the Baron's book so you may examine the diagrams and directions for yourself?"

This time Aleck had no words, but he clutched the precious book tightly all the way home, and then promptly sat down, as he said later, "to devour it." It was in French, but Aleck's education had included enough French for him to understand and enjoy it, and as he read his idea grew and strengthened. When he'd finished the last pages, he sat still for a brief time, with ideas swarming in his head and waves of eagerness sweeping over him, and then he stood up and went in search of Melville.

"Melly," he announced, "this is a very exciting book. Sir Charles used it to build his 'speaking machine.' I think we could make one just as good—perhaps a little better."

The book captivated Melly too. He was enthusiastic, and when they took the idea to their father, Melville Bell was delighted. Although both boys were mechanically inclined, this certainly would be their most ambitious attempt, and their father was proud of them. His encouragement was generous, and he ended by giving them some advice. "Don't think you have to copy either Baron von Kempelen or Sir Charles. You know what vocal organs are like. Copy nature. You can do it, boys. You can do it!"

It was Aleck who mapped out their program. "I'll make the head and mouth, Melly, and you can do the throat. We'll begin by making a gutta-percha cast of the skull Papa gave me."

It really was a wonderful contrivance. Aleck covered the head with a "skin" of soft rubber stuffed with cotton batting to form the cheeks. The lips were made of a frame of wire covered with rubber, and there was a piece of wood hinged to what was supposed to be the palate. Melville made the throat of tin and the windpipe of rubber tubing. They had meant to contrive a tongue, but they were dancing with impatience to assemble the speaking machine and try it out. They fastened the two sections together, and Melville blew into the rubber tubing. Out of the mouth came a strangely human sound as if someone were singing "ah-h-h!" Aleck opened and closed the rubber lips hurriedly while Melly blew again, and suddenly the speaking machine was saying very clearly, "Mam*ma!* Mam*ma!*"

The boys looked at each other and a beautiful idea flashed between them.

"The stair landing," breathed Aleck.

"Let's!" agreed Melly enthusiastically.

Gleefully they carried the speaking machine to the door and opened it carefully. No one was in sight. This was wonderful. They stole out upon the stair landing that led to the various flats, and then they let go! Melly took deep breaths and blew gustily into the rubber windpipe while Aleck manipulated the lips vigorously, and soon the hall was "resounding," as Aleck put it later, "with the most agonizing cries of 'Mam*ma!* Mam*ma!* Mam*ma!*' It really sounded like a baby in great distress!"

Suddenly a door opened upstairs, and a woman cried out in dismay, "My goodness, what's the matter with that baby?"

She never found out. Clasping their hands tightly over their mouths to control their giggles, the two boys stole quietly back to their own door. Their Speaking Machine certainly worked!

~ 6

The Turning Road

AND NOW ALECK'S LIFE TOOK ITS FIRST SHARP TURN.

For some time he had been heading for disaster in his school. Even a promising young scientist who deciphered Visible Speech at sight and could help build a "speaking machine" could acquire low grades if he blithely persisted in ignoring all the subjects the Royal High School counted important, and Aleck Bell's grades were perilously near rock bottom.

"I do enjoy music and botany and natural history and science," he had assured his parents when he was faced with that unpleasant fact, "but they aren't considered at the Royal High School, and Latin and Greek just aren't to my taste!"

"Oh?" And geography—or mathematics?" his father prodded.

This was not the same father who had been so happily proud of the speaking-machine and the rest of his son's interests, but a person with troubled eyes, and Aleck was suddenly ashamed. "I—I just can't get the right answers in arithmetic," he blurted reluctantly, "and geography is—boring."

A few weeks later his mother was packing a trunk and events were moving swiftly. Aleck found himself aboard a London-bound train. He was to go to his Grandfather Bell's in Harrington Square for a year's stay. He had been thrilled by the prospect when he had first heard of it, but now, watching the Scottish scenery fall behind the train windows, he was beginning to feel—well, uncertain. London might be exciting, but he would be separated from Melly and their inventions for a year—and Aleck had a sudden suspicion that inventions and experiments might not be welcomed at 18 Harrington Square, or indeed that he'd have time for them. Grandfather Bell had been headmaster of a school in Dundee, and Aleck could guess the reason for his "invitation" to his grandson just at this time.

The house he finally reached was a high, austere-looking place seen through the curling yellow fog, and when the door of Number 18 opened and he was bidden to enter, something about the whole atmosphere made the boy feel as if he had to hold his breath as he followed the servant to the drawing room where his grandfather was waiting.

"Master Alexander, sir."

There was a fire on the hearth, but the draperies were heavy and the furniture dark, and somehow the whole house had a lonely atmosphere, painfully different from the cheerful, lively cottage in Edinburgh. But the man sitting by the fire had a warm smile and his voice was friendly. "Come here, boy, and let me see what you are like, Alexander Bell the third."

When his first wave of shyness and nervousness had passed, the boy saw that Alexander Bell the first was a decidedly handsome and distinguished-looking man with beautiful, snowy, thick hair like a halo around his head; penetrating, dark, deep-set eyes and expressive hands. He was seventy-one now, but he was still as erect and lithe as a young man. The thing that attracted his grandson, however, was the sound of his voice, vibrant and clear, like the notes of a wonderfully-tuned bell.

For his part Alexander Bell was seeing a tall, thin, rather pale and gawky youngster with intelligent eyes and tousled black hair, and clothes that made the fastidious gentleman shudder.

63

"Sit down, Aleck, and I'll ring for tea. You must be hungry. Now the first thing to do for you is to get you into some decent clothes," Grandfather Bell commented after he had finished his first appraising scrutiny.

Aleck suddenly felt hot and awkward and unkempt. He flushed. "Papa and Mama thought these clothes would be suitable for a schoolboy, sir."

Probably his grandfather saw how miserable he was, for his smile and voice were suddenly kind. "Undoubtedly in your Edinburgh high school your tweeds did very well, but you are going to live in London now, and if you are going out as my grandson you must dress like other boys of your position. I will send for my tailor in the morning."

In the next few days it began to seem to Aleck that there was very little that Grandfather Bell did approve of about his namesake. His probing questions left the boy feeling guiltily careless and lazy, and his grandfather's verdict did nothing to alter the feeling. "You are grossly ignorant of the ordinary subjects that every schoolboy should know. I doubt if you are a dullard, however, for your father thinks well of your abilities in some other directions, and if you really desire to remedy the defects in your education, I have every reason to believe you can, and I propose instructing you in elocution and English literature myself."

Alexander Bell had been several things since he had abandoned his shoemaker's bench in his youth: actor, teacher, headmaster, dramatic reader of Shakespeare's plays—and finally, the real pioneer in the scientific study and correction of speech difficulties, a "corrector of defective utterance." But he was first and always a gentleman—more thoroughly a gentleman (although his grandson couldn't know it at the time) than many Londoners who could claim titles.

Above all else he was a real teacher, one who could make learning whatever he was teaching exciting and fascinating; and Aleck hadn't been his pupil more than a few days before he was ashamed of the lazy, careless record he had made at the Royal High School. But nobody in Edinburgh could ever have made English literature the dramatic, glorious thing that was coming alive under his grandfather's teaching, or taught him to do all the marvelous things Alexander Bell the first could do with his voice. The house might be touched with loneliness—the second Mrs. Bell had died the year before—and Aleck himself might be almost engulfed in homesickness at times, but something was stirring within him that he had never known before.

In one thing Aleck did surprise his grandfather—his flair for music. Aleck couldn't remember a time when he and the piano were not well acquainted. He could play anything he heard by ear, and improvise by the hour, a gift he inherited from his musical and artistic mother. Even in Harrington Square, with his aristocratic grandfather as audience, once he touched the keys of the fine piano, something happened. He wasn't an uncomfortable, awkward schoolboy on the piano bench; he and the piano belonged together. He played scraps of old songs, parts of melodies he remembered having heard his mother play, and a few ideas straying through his own head, weaving them all together until he forgot the awesome house and his grandfather's ability to make him feel insignificant, and was only conscious of the way the piano seemed to answer him and the surge and sweep and lilt of the music.

But answering his grandfather's questions afterward, the boy felt a little shamefaced. No sir, he hadn't had music lessons—his mother had shown him a bit about the piano, but that was all. No, he couldn't read music; he just played what he remembered or what came into his head.

Uncomfortably aware of his grandfather's low opinion of most of his talents, Aleck must have been surprised when he was informed that he was to see and play for someone Grandfather called "a distinguished professor of music," Signor Auguste Benoit Bertini, whose judgment of Aleck's musical ability could be depended upon. Perhaps he went to the appointment a little nervously, but Signor Bertini was

65

a kindly, friendly man, and before he knew it Aleck was improvising for him as easily and freely as he would at home. Both the Bells must have been startled by the professor's enthusiastic outburst.

"This boy has had no musical training except a few scattered lessons from the mother? He knows nothing of reading music? Teach him? Certainly, yes, yes! I shall make of him my special protégé. I will teach him my very special system —train him as my successor!"

Whatever Alexander Bell thought of this prospect, he made no objections. Perhaps he acknowledged that here was something about which Aleck was neither indifferent or indolent.

At any rate the hours he spent with Signor Bertini became the happiest Aleck ever spent in London. He learned everything the professor taught easily and joyously, and when the professor talked jubilantly about his career as a professor of music and a concert pianist, Aleck knew he had a new ambition.

But Signor Bertini was an old and ailing man. He and Aleck had only a matter of months together until the day when they both knew they had come to their last lesson, when the old musician paid him the real tribute of piling all his notebooks and everything else connected with teaching his famous "system" into the arms of the fourteen-year-old boy and saying, "Take them. Take them. And when you are grown up, go on with my work. Don't let my name be forgotten!"

Aleck never had another teacher of music. Perhaps he had no time. "I will help you to map out your time and show you how to devote certain hours to the study of your school subjects," his grandfather had promised. And he did.

"He made me ashamed of my ignorance," Aleck said years later. "This year with my grandfather converted me from an ignorant and careless boy into a rather studious youth anxious to fit himself for college."

There was really nothing else for the boy to do except

spend most of his time studying or reading. He and his grandfather lived alone except for the servants, and his grandfather was busy with his speech classes. Aleck had no companions of his own age in London. As for recreation —well, there was the small enclosed park on Harrington Square. Aleck had been bewildered and astonished when he first came and his grandfather produced a key for him. "To the gate of the park. Also be sure to lock the gate. This is a very private park, just for the residents of Harrington Square."

When he did walk there—or anywhere else in London— he was expected to wear the fashionable clothes the tailor had provided: skin-tight trousers, short Eton jacket, and— crowning glory—a top hat! He thought longingly of his Scottish tweeds, and acquired a permanent distaste for uncomfortable clothes. But—worst of all for an active boy who had known the freedom of the hills—"I had to carry a *cane!*" he told his family in disgust.

He was also obliged to memorize long pages from *Macbeth, Hamlet, The Merchant of Venice* and other favorites of his grandfather's, and recite them until Alexander Bell was satisfied. He was "permitted" to observe his grandfather's classes so that he could study his methods.

By the time he returned to Edinburgh he was a little thinner and paler than when he left his family, but he wasn't ignorant. He also discovered that he wasn't satisfied with Edinburgh, and with being his father's assistant in demonstrating Visible Speech. The year in London had made him grow up, and he "resented being treated like a schoolboy." He fleetingly considered running away to sea, but that was a little too drastic. He and eighteen-year-old Melly decided to watch the "Wanted" column in the newspaper *The Scotsman.* Two weeks later Aleck gleefully spotted an advertisement.

"Look, Melly, this would be exactly the right thing! 'Mr. Skinner, Headmaster of Weston House, in Elgin, Moreyshire, desires two pupil teachers, one for elocution, the second for music!' It couldn't be better! You can apply for the elocution, and I for the music position!"

Probably Melville was yearning for a little independence and experience beyond their father's circle himself. At any rate he was enthusiastic, and proceeded to write a very persuasive letter, forgetting to say that Alexander Graham Bell, applicant for the post of music instructor, was just sixteen, and Melville James Bell eighteen. Naïvely enough, they gave their own father's name as a reference! He promptly heard from Mr. Skinner, much to the confusion of his sons, but the Bells must have been extraordinary parents, for the affair ended with Melville's being engaged as his father's paid professional assistant, and a very surprised Aleck's going to Weston House to fill both positions.

He lived a crowded life for the next few years, alternating between teaching and attending the Universities of Edinburgh and London—and sometimes doing both together. He taught at Weston House, and at a little school for the deaf near London. Grandfather Bell died five years after Aleck had left him, and leaving Melville in charge of the Edinburgh Bell practice, the boys' father went up to London to take over his father's work.

Two years later tragedy had come to the Bells. Edward Bell, who had never been robust and had grown too rapidly for his slender strength, fell a victim to tuberculosis at eighteen. Three years afterward Aleck watched with horror as the same scourge struck and claimed his beloved Melville.

And then the older Bells realized with sickening suddenness that the dark shadow was hovering over Aleck. Perhaps, rather than any definite illness, his trouble was caused by the double load of teaching and work at the universities, and after that the task of acting as his father's assistant and being in complete charge of the Bell practice of Corrective Utterance in London while his father was lecturing in Boston, coupled with the work of nursing his brothers and the grief of their loss. Aleck had frightening attacks of exhaustion and periods of being feverish, and one specialist warned that he had only six months to live.

Alexander Melville and Eliza Grace Bell did not hesitate. They did not intend to let the fog-filled London atmosphere or the bleak climate of Scotland rob them of their last son. The Bells were justifiably proud of their work. They were not quacks, nor did they offer fads or untested theories. They could teach elocution and cure or control speech troubles because they had made a scientific study of human vocal organs, and of sound and speech. But Melville Bell sold his practice and everything else, and took Aleck and Melly's young widow, Carrie, to Canada. He had spent part of his youth in Newfoundland and had visited Canada then.

He found a home for his family on Tuleto Heights, four miles from the town of Brantford. "I have seen no place in Ontario," he told them, "that I would prefer for a pleasant, quiet and healthful residence."

It was pleasant. The house was small, but it had an orchard and a lovely grove of slender birch trees, and the Grand River flowed past a bend at the foot of the property. Aleck lay in a hammock under the trees, or took pillows and blankets to a nook above the river that he called "my dreaming place."

For a few months the Bells must have waited with bated breath and prayers on their lips. And then the clear, unimpeded sunshine and the pure clean Canadian air worked their miracle. The shadow lifted from Aleck Bell, never to return. He was still too thin and too pale, but he was well, and the strength that had been wasted in his youth was beginning to flow back.

His thoughts, as he lay in his "dreaming place," couldn't have been very bright. He must have wondered if his life was over at twenty-three. What was the use of his scientific training? His careful experiments in sound? The years he had spent in teaching—particularly the tests he had made in using Visible Speech to teach deaf children to speak?

Canada was still a rural country. What use would any of the things he knew be here?

Then Mr. Bell went on a lecture tour to Boston in the United States. And when he returned he had a proposition for his son.

The Boston Day School for the Deaf had just been organized and the principal, Miss Sarah Fuller, was very greatly interested in Visible Speech. She had asked him to consider coming to Boston to give the teachers in her school a course in the new subject and how to use it for the benefit of the deaf.

"I told her," Melville Bell reported to his family, "that I had spent twenty years in devising the system, and now I didn't wish to teach it, but that I had a son who could come to Boston if it seemed advisable to take the risk of affecting his health by the change of climate." He repeated all this slowly and patiently into his wife's ear trumpet, and at the sight of the quick anxiety in her face he patted her hand reassuringly. "Miss Fuller told me that whatever she could do to care for and protect my son's health would gladly be done, and that she had a sister who would welcome him into her home at any time."

So in April, 1871, Aleck Bell, at twenty-four, took another turn in his road and came to Boston, where he would meet a fifteen-and-a-half-year-old girl he hoped would become the balance wheel of his life.

7

The Talking Glove

IF MABEL HUBBARD COULD HAVE KNOWN OF ALECK BELL'S FIRST
impressions of the teachers at the Boston Day School for the
Deaf, the unflattering opinion she had of him two years
later would have turned to positive distaste, because for
some reason he did not admire Mary True or a couple of the
others at their first meeting.

Sarah Fuller, however, captured his heart at once. "I
never saw *love, goodness and firmness* so blended in one
face before," he exclaimed to his parents and Carrie on the
sixteenth of April. "Her abilities are very great, and she is
overflowing with genuine goodness toward the children."

Miss Fuller's goodness and generosity went out to the
young Scot as well, even as she had promised his father.
Aleck had found a room in the boardinghouse where his
father had lodged during the lecture tour, but Sarah Fuller
and her sister Mrs. Jordan opened both their doors and
their hearts to him, and he came to them as often as he
could, responding to the warmth of the friendliness and
understanding he found there. Sarah Fuller observed that
"if he had been Mrs. Jordan's own son he couldn't have received more love and sympathy during his first years in
Boston."

This certainly seems to have been the truth, and she
might have added that she herself gave her young visiting
"Professor Bell" her full share of interest and love and
understanding.

71

"He often came to our house in the evening," she commented, "and worked on his experiments (he was always experimenting with means to transmit sound) until past midnight—and sometimes until two in the morning!" This might have been trying to a busy principal who had to be in charge of a school the next day, but evidently Miss Fuller was just as engrossed in the proceedings as Aleck was, for she hastens to add, "I always sat with him and followed his successes and failures with extreme interest."

It wasn't difficult to be interested in Aleck Bell—even at two o'clock in the morning! He was the exact opposite of a traditional "dour, taciturn Scot" when he was kindling with an idea. The eager light in his face, the conviction that rang in his voice and his magnetic enthusiasm could be very infectious.

"Look, Miss Fuller!" he exclaimed one night, "I've brought this toy balloon. Now if a deaf child holds it tightly in his arms and against his chest like this, he would receive the vibrations of the noises around him. Wouldn't that be valuable in helping him to identify sounds?"

Sarah Fuller put down her mending to clasp the balloon while Aleck provided a series of test noises, repeating them until she dropped the balloon and nodded her quick approval, her own face alight.

"It might be especially useful to convey street noises to the deaf, particularly after dark when they need to be warned against approaching vehicles." The triumphant look faded from the intent face, and his voice was suddenly wistful. "If only I could devise a machine to write sound, just as it is spoken—a machine to hear for these children!"

Sarah Fuller did not laugh. Something in Aleck Bell's face told her that this young man wasn't just a bright young professor with a promising interest in science, and that what he had just said wasn't a passing fanciful idea. He might be passionately intent on going after what he wanted, and she had a flash of intuition that he would find it. Suddenly she gave him a look of sober understanding. "Someday you will invent something very wonderful," she told him. "You will be famous someday."

Boston was very good for young Professor A. Graham Bell. Behind him in England and Scotland had been illness and weakness, disappointment and bitter loss; but here in the New World health and strength had flooded back, and in this American city all roads lay open before him. Now he knew which one he wanted to take: he would be a teacher for the deaf.

His engagement with Miss Fuller's school lasted only a few weeks, but Visible Speech had been an enormous success and so had A. Graham Bell, with both the teachers and the children. Aleck reported to his family that "one sweet little girl, Josey Ward" had even prayed about his departure, "I'm sorry Mr. Bell is going away Wednesday. Aren't You? I think Mr. Bell will tell his mother—You know his mother, don't You, the lady with the long, long trumpet? Don't You think Mr. Bell will tell his mother about the deaf-and-dumb children?"

Other schools had been hearing of Professor A. Graham Bell and his phenomenal system. Before he left for Brantford the first of June he had a contract to teach and lecture at the Clarke School in Northampton, and probably had talked matters over with the school's president, a gentleman named Gardiner Greene Hubbard.

Back in Brantford Aleck entertained himself that vacation by becoming friends with the Mohawk Indians who lived on a reservation nearby. He was fascinated by their language and songs, and translated them into Visible Speech, and the Indians paid him the very rare compliment of adopting him into their tribe, even making him an honorary chief. Aleck chuckled about it in his diary:

August 13. Onwanonsyshon (Chief Johnson) took me into his private room and dressed me up in full Indian costume! I wore his buckskin coat with silver britches made by the Indians from silver coins—chiefly shillings,

73

half crowns and crowns. Upon the breast of the coat hung various medals—upon my head was placed his hat of eagles' wings and ostrich feathers—which brought my height up to nearly eight feet.

Dressed in full costume and with a tomahawk in my hand I stalked majestically into the room where all were seated—frightening my mother nearly out of her senses.

The Mohawks also delighted Aleck by teaching him their war dance, which he learned and danced enthusiastically, enjoying it so much that for years, whenever he was especially happy or excited, he would cavort in the Mohawk war dance, with its accompanying yells and other weird noises, much to the astonishment of anyone who happened to be nearby, and the dismay of various landladies.

Visible Speech was just as great a success at the Clarke School as it had been in Boston. Amazed teachers reported that "in a few weeks Professor Bell had taught the children to use more than four hundred English syllables, some of which they had failed to learn in the two or three years under other methods." From Clarke he went to the American Asylum in Hartford (what had become of Mr. Turner, who had so fervently announced that he hoped speech would never have a part in Hartford's teaching?) for a course of lectures and demonstrations.

There were several openings for A. Graham Bell now. There were rumors that when she returned from her year's leave of absence abroad, Harriet Rogers would resign as principal of Clarke School and the position would be offered to Professor Bell.

He asked his parents' advice about it. He could become a private teacher. Mr. Dudley, for example, had become so excited about Visible Speech that he wanted Theresa to have a course of private lessons, and Aleck's good Samaritans, Mrs. Jordan and Sarah Fuller, took the girl into their home, and another pupil, Alice Jennings, came as well; but Aleck had become dissatisfied and tired of being a wandering lecturer. He wanted to do something that would be permanent and established. He proved his admiration for Sarah Fuller by going to her for advice also, and she gave it promptly.

"I urge you to make Boston your place of residence, and advertise a plan for receiving pupils. Either teachers to be trained for work with deaf children, or deaf persons—adults or children. It seems to me that a fixed address will give dignity to your announcement and inspire confidence."

She even offered to go with him on the quest for rooms to use as his lodging and school, and found them at 35 West Newton Street. When the Conference of Principals of Institutions and Schools for the Deaf held its meeting in Flint, Michigan, in the summer, she saw to it that Professor A. Graham Bell received an invitation, and introduced him to everyone as "an exceptional man."

All of which proves that she herself was a most exceptional woman. A lesser one wouldn't have taken such risks with her own success, for here was a new teacher with a dynamic personality and an astonishing new system. If his proposed school was the breathtaking success his lectures had been, it might do drastic things to her own. But Sarah Fuller sensed that here was someone who was destined to do extraordinary things, and she went blithely ahead opening every door of opportunity she could find for Aleck Bell.

In the Fall when he returned to begin his new work in his new quarters, she introduced him to the first of the two men who would help him play the most important role of his life. "Mr. Bell, I have been telling Mr. Sanders about you. Mr. Sanders has a son born deaf. I have been giving him instructions in my home for the past three weeks, and he has made a most valuable beginning, but he is only five years old, and too young for my school. Will you consider taking him as a private pupil?"

"We would see that his nurse accompanies him," Mr. Sanders struck in eagerly. "I have made inquiries, and we can make arrangements for them to live in the same house where you have your rooms. Georgie's mother and I would consider it an honor if you will undertake his education. Miss Fuller speaks highly of your qualifications."

Aleck had never dealt with so small a child, but it would be a happy beginning for his new venture, and after he had seen the bright, lovable little chap, Aleck agreed with enthusiasm.

"October 2, 1872," he wrote home. "Georgie Sanders made a good beginning yesterday. He lives in the same house. I propose to divide his education into two branches—speech and mental development. . . . Mr. Dudley here today. Theresa Dudley is coming to take lessons for three weeks or a month. I hear Mabel Hubbard will be a pupil soon."

It was a good beginning. He shut up his inkwell and looked around the room with a quick thrill of pride. This room would be his classroom—his own classroom for his own school! He had been teaching since he was sixteen, but always in other teachers' establishments. Now, thanks to Miss Fuller's advice and kindness, he had his own, and he had pupils.

He looked around the room with satisfaction. It really was elegantly furnished, with its black walnut, hair-covered furniture and light-colored carpet. A stranger wouldn't guess that it doubled as his bedroom, with the sofa such a deception, folding over to form a bed. He could see through into the reception room with its green furniture and marble-topped table and its green and black carpet, all very fashionable. Yes, this was a good place to begin his career—not only to teach, but to work out his beloved experiments.

It was the beginning of a hectic year. Aleck taught; he lectured; he wrote magazine articles about Visible Speech; and he even started a small magazine of his own, "The Visible Speech Pioneer," which he had to copy in longhand to send to the various teachers he knew would be interested.

And of course there was his special pupil, little Georgie

Sanders. Aleck began with him by visiting the child and his nurse in their rooms, observing him, noticing that his special toys were a doll and a toy horse; he frolicked with the boy, played games with him, helped put the doll to bed, and even went through the action of solemnly giving it a drink of water, at which Georgie danced up and down, screaming with delight at the way this new playmate understood his world!

Aleck smiled at him, gave him a farewell hug and went back to his bed-classroom to evolve a way to begin the child's education. The toys—wouldn't they make a good start? If the boy could be shown how to associate words with the doll, the horse and his ball, wouldn't that do it?

When Aleck knocked at the door of the rooms occupied by Georgie and his nurse the next day, he was armed with a small blackboard and a white cardboard box. Seeing the tall stranger who had understood so well that a doll liked a drink at bedtime, Georgie darted over and hugged his knees. Then he spied the cardboard box in Aleck's hand and stood on tiptoe to peer curiously as Aleck, with an air of great mystery, raised the cover. Perhaps the child had expected candy, but the box contained small white squares of cardboard. He followed Aleck around the room, watching intently as this newcomer in his life pasted one of the squares on the doll's forehead, another on its bed, one on its little chair, and others on the horse and ball. Georgie had no way of knowing it, but these squares were the Open Sesame that would take him into the world of language, for each card bore the name of the object it adorned—"doll," "bed," "horse" and the rest. Aleck called his attention to the writing, showed him there were other cards in the box with the same kind of marks, and then, holding up the doll and the horse and the ball in turn, he wrote the words on the blackboard.

All his life Aleck enjoyed being dramatic, and his instinctive dramatic flair helped him enter Georgie's world. They played with the doll, giving it a ride on the horse, which of

77

course called for the word "ride," and fed it its supper, with the words "drink" and "cup," and put it to bed. Georgie was entranced, but Aleck was careful not to keep up the play so long that it would tire or confuse the child.

Georgie learned easily, and a few weeks later they were playing another "game." For this one, Aleck fitted on the boy's left hand a glove inked with the letters of the alphabet. By this time Georgie knew his letters and could correctly place the separate labels or even the individual letters on the doll or the bed when Aleck indicated the desired object. With the little boy wearing the glove, Aleck would touch each letter lightly to spell out a word. This would be their way of talking to each other—a way anyone could use. It seems a little strange that he didn't teach Georgie the finger alphabet generally used by deaf-mutes, which he certainly knew. Perhaps Aleck had decided that one alphabet, the one familiar to everybody, would mean less complication and confusion to a small boy, besides giving him a means of communication everyone could use, wherever he might be.

At any rate they both became so expert with "the talking glove" that Aleck could use all five fingers at once, much as though he were playing the piano, and Georgie learned to "read" his teacher's fingers almost as rapidly as a person's eyes follow a printed page. Sometimes he could even anticipate Aleck's meaning before the spelling was complete.

Now Georgie's lessons included reading and writing, and by the time he was six he was producing letters that would have been a credit to boys several years older.

Like the Hubbards and Mary True, Aleck refused to recognize gestures. Spelling and writing were doors open to Georgie—but not "signs" or pantomime. Later Aleck would teach him speech, but lip reading wasn't included in his plans. Oddly enough, despite his work with Sarah Fuller and at the Clarke School, Aleck was extremely skeptical about lip reading. He really considered it complete humbug. This was before he met Mabel Hubbard. For some reason

Mabel never did appear at 35 West Newton Street, although Aleck was probably much too busy these days to think or care about her failure to appear.

He really was burning his candle at both ends, and in the middle as well. Teaching Georgie Sanders could easily have been a full-time occupation, but Aleck had his hands full with his other pupils and his little magazine, and in addition he was beginning to work on an experiment that had been teasing his brain for some time—an idea for a "harmonic" or multiple telegraph.

To a Scot who was used to the speed and efficiency of the British telegraph system the slowness and uncertainty of the American telegraph companies was both amazing and irritating. If one telegraphed an urgent message it could dally along the wires and arrive much too late, and sometimes an ordinary letter would outrun a telegram.

Certainly the trouble didn't stem from any lack of wires; there were spider-webs of wires crisscrossing American cities now. The idea that had been teasing Aleck was simple: Why couldn't there be some way to transmit more than one message at a time? Why couldn't ten or a dozen messages be sent along the same wire simultaneously? Aleck Bell was never a man who merely wondered about a thing; he had to investigate all the possibilities for himself. So after his pupils had gone for the day, and he had finished with Georgie Sanders, he attacked the multiple telegraph.

There was an obliging Mr. Richards who had the room opposite Aleck's in the adjoining house, and he let him string all sorts of wires between the two rooms, and Aleck would work on and on, contriving instruments, testing "reeds" and "rods," making adjustments, ignoring time until the milk wagons came clattering along the streets and the fingers of the dawn warned him he was facing another day. He skimped on food to buy materials for the telegraph; he skimped on sleep to work on it. It was no wonder that by the time summer arrived he had squandered his rather slender supply of strength and was forced to admit to his

79

parents and Carrie that he was so nervously exhausted that he couldn't make coherent plans for his next term's work. On July first he was confessing

> As I have scarcely had any sound sleep for many days past on account of imaginary noises at night I thought it best to consult Dr. Cotting—who recommends giving up work as soon as possible. . . . He seems to think that my mind being so much bent on the study of noises all day long may have induced these effects.
>
> If I could only bring little Georgie with me to Canada, I would be content to relinquish everything else and go home 'right away.'
>
> It is no labor to teach him as my plan is to let him imbibe instruction through play.
>
> If you could accommodate myself and Georgie (with his nurse, Fredericka, to look after him) and if you approve of the plan let me know at once . . . If you look favorably upon the plan please telegraph to me before Saturday afternoon and I can broach the subject to Mr. and Mrs. Sanders—who would, of course, require a few days' preparation.
>
> I have come to the conclusion I have not the strength to do all that I want to do and I have resolved to rest next year and to take no pupils except those connected with the deaf and dumb.
>
> My plan is this—I can have free rooms and board with Mrs. Sanders of Salem (Georgie's grandma), in return for instructing him. Mrs. Sanders has been very kind to me and I could not have a more comfortable home than hers. Salem is only a few miles from Boston and I could come into town every day.

He told them that he had been offered the chair of Professor of Vocal Physiology in the School of Oratory at the new Boston University, and would receive a salary of five dollars an hour. He ended with the cry, "Please advise me!"

Whoever gave the advice, Aleck went to Brantford alone, and his mother used all her love and care to restore his health and strength to face another Boston winter. It would have been useless to ask him to promise to rest and sleep— at this period Aleck was too taut to be able to do either— but it must have been a comfort to Eliza Bell to know he would be in the hands of a thoughtful grandmother.

When Aleck reached Mrs. Sanders' comfortable and roomy century-old white clapboard green-shuttered New England house set in a wonderful tree-covered lawn, he had to acknowledge that this was a far more homelike and happy situation than any Boston boardinghouse could ever hope to be. It was to be home for Aleck Bell for well over two years.

And kindly, warmhearted Mrs. Sanders must have endeared herself to him forever when she smiled at him and remarked, "My son has been telling me of your interesting experiments, Mr. Bell, and I'm certain you must require a great deal of space for them; therefore I am going to give you the free use of the entire cellar for your workshop." (Eventually she was to give him the whole third floor of the house as well.)

That particular autumn and winter must have been a happy one in Aleck's life, for his classes were a success, and he could work on his telegraph in peace and comfort. Also, Georgie was forging ahead and "imbibing instruction" amazingly well.

The comradeship between the eager child and his tall young teacher was something neither of them ever forgot. Georgie was delighted when he understood that he and his adored Mr. Bell would be staying with his beloved grandmother in her big house—no more stuffy, small boardinghouse rooms to cramp an active little boy! Of course he knew Mr. Bell had to go to his school in Boston just the same as before, but when he returned there were the glorious times when Georgie had him all to himself, and they romped and played, and Mr. Bell wrote stories and drew funny pictures on the blackboard, and played games to teach him to talk like other people.

When the late afternoon's slanting sunlight over the lawn told him it was nearly time for Mr. Bell to return on the horsecar, Georgie would stand with his nose pressed tight against the windowpane, his intent eyes watching the street for the first sight of the familiar, lithe figure. He would wait until Mr. Bell waved his hat and then snatch up his "talking glove" and dart for the door. Most of the time Aleck would hardly be out of his coat before Georgie had catapulted upon him, breathlessly spelling out his doings for the day, then leaning against his knee to demand, "What did you see today? Tell me, please!"

Aleck was always ready. He had a genius for making the most ordinary sights fascinating and dramatic. "Well, there was a parrot riding on the car this morning. Yes, a real, gorgeous parrot, green and yellow! He looked so proud."

"Was he alone?" the little fingers hurriedly picked out the letters as the child's eyes widened.

"Oh, no, a lady had him in a big cage. And then I saw a monkey in a little coat on Tremont Street, and after that I spoke to a little dog, and I think he liked me, because he walked along with me for three blocks!"

The boy wiggled delightedly. "Make me a story about them, please."

Aleck would go to the blackboard and write the adventures of the parrot, the monkey and the little dog in simple words until his arm grew numb. Then Georgie would produce his book—the book that Aleck had made for him by writing out the child's favorite stories, using words of varying sizes to teach him the emphasis he couldn't hear.

"Will you write it here?" And then the child stood by, following the movements of Aleck's pen with eager attention as the little story was inscribed in the precious book.

There were other activities, too. Other children would have known they were "lessons"; Georgie Sanders thought of them as games, and anything introduced by Mr. Bell he entered into gleefully, even the difficult and seemingly senseless idea of learning to say words.

Georgie loved going to various places with his teacher, especially when they went to a place Mr. Bell called "Barnum's Circus." The great lions appealed to him particularly, and Mr. Bell held him close enough so he could feel them roar.

After Georgie's bedtime, and with his dinner finished, Aleck would hurry to his cellar retreat and plunge into work on the Multiple Telegraph. He worked too long and too late on it to please Mrs. Sanders, who was worried at his haggard face and frequent headaches, but she learned it was useless to protest. Professor Bell had every intention of discovering a way of sending multiple messages on a single wire simultaneously, and the only thing that would stop him would be finding the way! Mrs. Sanders shrugged and left him to his chaos of "reeds" and "rods" and "batteries" and "cells" and wires.

Oh yes, it was a happy era in Aleck Bell's life, and it was to prove better than he dreamed, for it was that fall that Mary True brought a former pupil of hers to his class in the basement room of the Boston University, and on that glorious day, Professor A. Graham Bell and Mabel Hubbard met each other.

∽ 8

"I Can Telegraph Speech"

BEING A VERY NORMAL GIRL, MABEL MUST HAVE BEEN PIQUED at having been dropped from Professor Bell's own personal class. Despite her first scornful impression of him she had reached the point where she was obliged to admit she enjoyed his teaching, and now—to be turned over to his assistant, Miss Locke, for no accountable reason! He had assured her mother she was a promising pupil, and he had no fault to find with her—and then he was silent. He had always been so pleasant and friendly; why had he turned so obdurate and arbitrary?

"Perhaps he has suddenly discovered he dislikes you," Berta teased her. She sat watching her sister's face with laughter-filled eyes. "Turn about is fair enough, you know. Perhaps Professor Bell simply can't stand you any longer!"

May flushed, then shrugged and brushed a strand of her shining, waist-length brown hair more vigorously than before.

"Well, he's rid of me if he wants to be! And if he won't teach me himself, possibly Papa won't send me to his classes!"

But if May had imagined that ceasing to be Professor Bell's pupil meant leaving the Visible Speech classes or ceasing to see him, she was doubly mistaken. Her parents were anxious that she continue working on Visible Speech, even though it would be with Miss Locke; and apparently it was Professor Bell's habit to stroll in upon his assistant's sessions with more or less regularity. And May was discovering something else: While Miss Locke might be a pleasant enough person, she missed Professor Bell's ingenious teaching methods, and his winning laughter when he had challenged her to produce the sounds he had pictured.

The day she braved a blinding snowstorm to attend class, and Mr. Bell insisted upon accompanying her home, she changed her mind slightly about his not being a gentleman. None of Sister's dapper beaux would have shown half as much solicitude for her comfort as Mr. Bell was showing her!

It was unfortunate for Aleck that the Hubbard family was a divided one that winter. Mrs. Hubbard and Grace and Berta were in New York staying with invalid Grandmama McCurdy, and Mr. Hubbard was trying important cases in Washington, so May was under the guardianship of her Cousin (always written with a capital) Mary Blatchford. While Cousin Mary frankly adored the girl, she could be a rather acid spinster toward anyone else, and it would be several years before she would regard Aleck Bell with anything but suspicious hostility. Tonight, as she opened the door to the two laughing, snow-covered figures, she gave poor Mr. Bell a look that must have been as chilling as the stormy air! Her thanks were prim, and when he said that he must be hurrying back to Salem, she agreed frigidly. "Yes, I expect it will be wise not to delay, so neither Mabel nor I will detain you!" and shut the door energetically as Aleck hurriedly bowed himself out into the swirling storm.

But when the Hubbard family returned to Brattle Street in the spring, the tide turned for Aleck. Gardiner Hubbard had met him several times, probably over matters concerning the Clarke School, and liked him. He had several invitations to tea, and then became a regular and very welcome guest at the Hubbard table for the Sunday dinner of roast beef and floating island pudding. Musical Mrs. Hubbard enjoyed the way he played the piano for her, and the young Hubbards and their relatives—the hospitable Hubbard household was always overflowing with guests: Lina and Augusta McCurdy and a youthful Hubbard cousin, Willie—were all fascinated by the experiments he persuaded them to take part in, like standing on the garden paths while he blindfolded them and then made tests to see if they could identify various sounds and the directions they came from. And they always crowded around him for his famous, shivery ghost stories. "Although I can never sleep at night after hearing them," Lina McCurdy would declare.

Lina and Berta and Grace united in teasing him unmercifully. They discovered one day that innocent Mr. Bell had never tasted that delectable American form of hot bread, waffles, and they proceeded to advise him enthusiastically how to get the most enjoyment out of them. "Pour plenty of oil and vinegar over them, Mr. Bell," Berta declared. "Oh, you certainly must try them that way! They are absolutely perfect!" No doubt Aleck thought Americans had most peculiar taste, but he might have screwed up his courage and tried the horrible experiment if Mrs. Hubbard had not come to his rescue.

It was Berta who set him to climbing after the highest cherries on the trees in the garden—those on the lower branches wouldn't do—and then induced him to sample the green gooseberries to see if they were ripe enough for her and Grace to eat!

Aleck responded to all the liveliness and foolery around the Hubbard house with alacrity. This was the first American home where he had been welcomed just as Graham Bell, and not because he was an expert on Visible Speech, and it was a very pleasant experience. He would never be a staid, woolgathering professor lost to everything except his teaching and his scientific investigations; and even though he was concentrating on his multiple telegraph with every fiber of his being, he was still the young man who delighted in whirling and stamping and yelling his way through an Indian war dance, and he certainly had had moments when he yearned for the company of some persons between the ages of six-year-old Georgie and the boy's grandmother!

Beside the pleasant, welcoming atmosphere of the Brattle Street house and the lively companionship he found there, there was Mabel and the chance to watch her, to get to know her and sometimes to talk to her—all very unobtrusively and discreetly. He prided himself he was being discreetness itself, but nothing escaped May's alert younger sisters.

"Call me pet names, dearie!" they teased her, "Call me 'Aleck!' "

Mabel flushed and protested they were being utterly outrageous. "Call Mr. Bell 'Aleck'? Why, he is almost an old man!"

Nevertheless it gave her a very pleasant sensation to have Mr. Bell seek her out for little chats by themselves. Even now, when her lip-reading ability had progressed to a point where she could readily understand practically everyone she met, many people outside of her own circle of family and friends made an embarrassing situation out of talking to her. They stumbled into the pitfall of too-slow and exaggerated lip motions, which were difficult to follow, or exploded one or two words at a time and left the rest of their sentences dangling, giving May the feeling she was groping for the rest. Most disconcerting of all, they would often answer her by addressing one of her sisters.

But with Mr. Bell it was different. They talked together as easily as though they had been doing it all their lives. Their favorite retreat was the conservatory, and there in the fern-filled, flower-fragrant room, Mr. Bell talked on topics that were truly amazing. He led her into political discussions and talked to her about his astonishing belief that women, as well as men, should take part in governing a country, and possess the right to vote. The girl felt as though Mr. Bell was opening new windows in her mind. And Mabel Hubbard had a very good mind. Years later one of her family was to call Mabel "a creative listener."

Mr. Bell talked about lighter things too, like the time

Prince Lucien Bonaparte had invited Alexander Melville Bell to dinner:

"Prince Lucien was a distinguished scientific man who made personal tours of Scotland, mapping out the geographical boundaries of the various Scottish dialects. As my father was a recognized authority on dialects, the Prince invited him to dinner to talk over the subject, and I also was included in the invitation. I was only a boy at the time, but I was duly impressed with the honor of dining with a real live prince. I did not understand very much of the conversation, and was more impressed, I think, by the elegance of the three waiters who stood at attention behind our three chairs. One put a plate with something on it right in front of me; and I was especially interested in the mysterious appearance of a hand the moment I let my knife or fork rest on my plate, followed by the sudden disappearance of the plate and the arrival of another!"

He paused to chuckle over the recollection. "I amused myself by counting the number of courses until I finally lost count. My boyhood memory was that there were over twenty, but I am a little more doubtful about that now!"

He also told her the story of a deaf child's version of the story of Job. A minister visiting a school for the deaf had written the story in graphic sentences on the blackboard and when he made his second visit he hopefully inquired if any pupil could tell him what God had done for Job.

"One small boy could." Mr. Bell chuckled. "He said 'God boiled Job seven days!' "

As the Sunday visits went on, Mr. Bell told Mabel of his own experiences. He told about Georgie Sanders, and the "talking glove" he had devised, and how the child was beginning to talk; and Mabel listened, her face alight, her eyes eager. But she was curious and puzzled about the use and need of the "talking glove." Her parents and Miss True would never have permitted her to possess such a thing!

Mr. Bell flushed and looked uncomfortable, and then confessed, "When I came to this country I had a little suspicion of Miss Fuller . . . I knew something of the mechanism of speech, and therefore knew that every deaf child could be taught to speak, but I thought it was humbug about the oral school. I thought their claims regarding lip reading were too big. That was my feeling. I believed in teaching speech, but I didn't believe in speech-reading." He looked at her with miserable eyes. "But I am converted. You did it!"

For the sake of that last admission May must have forgiven him the error of his early ways.

Did Aleck tell her of his Multiple Telegraph? and did she urge him to tell her father? At any rate, one Sunday he suddenly paused in the midst of his piano playing, swung around on the bench and looked toward the corner where Mr. Hubbard was reading. "Mr. Hubbard, sir, do you know that if I depress the *forte* pedal and sing 'do' into the piano, the proper note will answer me? Like this?" He pressed the pedal, bent forward, singing "Do—" and like an echo the piano responded. Mr. Hubbard put down his book and came toward the piano.

"And here's more!" Aleck was continuing eagerly, "if two pianos in two different places were connected by a wire, and a note was struck on one, the same note would respond on the other!"

"And what value is there in that fact?" Mr. Hubbard sounded perplexed but interested.

Flushed with eagerness now, Aleck explained his ideas for the multiple telegraph, the experiments he had been making at Mrs. Sanders' home in Salem, and the encouraging results he'd had. As he talked, an answering gleam sprang into his listener's face. Aleck happened to be bringing his inspiration to exactly the right person. For several years Gardiner Hubbard himself had been dissatisfied with the telegraphs and had been advocating improvements in the systems and trying to discover how they could be accomplished, and now here in his own music room was Mr. Bell with what promised to be a brilliant idea. Watching him intently, Mabel knew her father was growing enthusiastic.

89

"Mr. Bell, I believe you have a very sound idea there, and I am willing and ready to finance it and assist you in taking out the necessary patents."

Aleck's dark eyes kindled. "I would be very grateful, sir, but the fact is that just a few days ago I spoke to Mr. Sanders about the project, and he has very kindly offered to finance the work and help me obtain the patent for one-half interest in the completed telegraph."

Mr. Hubbard nodded his approval. "I think it would be excellent for you to have more than one backer. Suppose you speak to Mr. Sanders and ask if he would be interested in dividing the ownership three ways?"

Thomas Sanders was a well-to-do merchant with a prosperous leather business. By now, having watched his son emerge from a pathetic state of dumbness, in the starkest sense of the word, he was prepared to believe that A. Graham Bell could make anything break forth into sound. He had seen the awe-inspiring array of wires his mother had indulgently let Mr. Bell string all over her house, and he had heard the signals Mr. Bell coaxed from them. Not that he understood the theory of the thing—but Mr. Bell believed in it, and Thomas Sanders believed in Graham Bell!

He would have been glad to sponsor the undertaking alone, but something in Aleck Bell's face when he spoke of Mr. Hubbard's offer told him that for some reason Aleck really wanted Gardiner Hubbard included, so Thomas Sanders good-naturedly agreed to have Mr. Hubbard join forces with them, and to give him a one-third interest in the Bell inventions. This arrangement was to prove fortunate to everyone concerned, for before they could reap any reward for their investment in Aleck Bell, Mr. Hubbard's resources were to reach a perilously low ebb, and Mr. Sanders was to face a debt of one hundred and ten thousand dollars.

Now that he had financial assets there was one bugaboo that disappeared from Aleck's life forever: He would no longer have to make his own apparatus! He had always been

able to imagine, plan and sketch all sorts of working con-
trivances, but when the time came for actual construction, he
had been in misery. Melly had been "neat-handed," but
Aleck was awkwardness itself, pathetically clumsy and heart-
breakingly slow. In the early stages of developing the multiple telegraph he had struggled to make his instruments
himself, although he knew only too well he had neither the
skill nor the knowledge to do it properly. Now, however, he
investigated at once when someone suggested that he try the
Charles Williams Electrical Shop in Boston. The shop must
have seemed like a paradise to Aleck, for here was a corps
of skilled young electricians who would try to produce any
device a would-be inventor desired. Aleck was delighted at
the prospect, and promptly left specifications for a transmitter and a receiver to be used with his multiple telegraph.

However, when he went to claim the finished products his
eager eyes spotted a flaw in them. Mr. Williams apologized
immediately. "Tom Watson made them, and he's one of my
best—the young fellow over there. I'll speak to him at once
and—" Before he finished, Aleck was striding over to Tom
Watson's bench.

Twenty-year-old Thomas Watson looked up to see the
tall, thin, "quick-motioned man with a big nose and high-
sloping forehead" and intense eyes hurrying toward him. He
was holding out the transmitter Tom had made, and the
young man braced himself against a torrent of reproaches.
Tom was accustomed to dealing with aggrieved inventors
when their precious devices went wrong, but this Mr. Bell
was explaining quietly and clearly what he wanted changed,
and why. What struck the young electrician most of all was
his courtesy in doing it. That, and the beautiful simplicity
of the transmitter. Thomas Watson had seen some weird
contraptions emerge from the specifications brought into the
shop, but this looked as if it might have possibilities.

Thomas himself believed that the Williams patrons liked
him because he was a rapid worker. But it was more than
mere deftness and speed that attracted Aleck Bell to this

91

smooth-shaven, intelligent-faced youngster and eventually prompted him to ask young Mr. Watson if he would like to become his assistant in his telegraph inventions.

Mr. Watson would, indeed! He was fascinated with Mr. Bell's ideas, thrilled and honored to be working with him, and captivated by the man himself. "No finer influence than Mr. Bell ever came into my life," he said afterward.

By a coincidence young Thomas Watson also lived in Salem, and once he and Aleck Bell were friends they were often together. When Aleck invited him to dinner, the embarrassed young man found himself involved with a number of forks he didn't know how to handle except by keeping an intent eye on his host and following his example. Afterward he listened with astonished delight while Aleck played the piano, and stared in awe at his collection of scientific books. When Tom shamefacedly confessed his ignorance of science and the fact that he had dropped algebra, the teacher in Aleck came to the fore and he introduced Tom to the best scientific authors and bought him the latest book on algebra.

They formed the habit of taking long walks together, and it was during one of these walks that Aleck made a remark that strained Thomas Watson's faith in his judgment almost to the breaking point. They had been striding along the beach at Swampscott with the curling waves almost at their feet and Bell's eyes intent on the graceful, whirling gyrations of the sea gulls, when he suddenly remarked, "You know, Mr. Watson, someday we will be building machines which will be capable of flying through the air like these gulls, only at a far greater speed!"

Thomas threw him a startled look to see if this might be a tongue-in-cheek remark, but from the flaming eagerness in the other's eyes, he knew it wasn't.

"I have been thinking," Aleck was continuing with the eagerness that always flashed in him at the dawn of a scientific idea, "that it might be powered with that light steam engine and boiler I saw you working on the other day. Only it would naturally have to be on a far larger scale."

When Aleck was aflame with enthusiasm his conviction was contagious. Young Tom Watson began tingling with it himself, even though his reason told him that the idea of "machines" burdened with engines winging their way through the skies was utterly absurd and fantastic. It was exciting, and as Aleck proceeded to unfold his ideas for his "flying machines" they almost seemed believable. Thomas Watson suspected that if Mr. Bell hadn't been pledged to work on the multiple or "harmonic" telegraph, they would have plunged into the study of flying machines immediately.

As it was, they had a sufficient number of inventions to keep them desperately busy. Besides the multiple telegraph, Aleck was contriving an "autograph telegraph" that would send telegrams in a person's own handwriting, and trying to perfect the invention closest to his heart, the "machine to hear for deaf children." This would be the "phonautograph," or "sound writer." They had something of the sort at the infant Massachusetts Institute of Technology, and Aleck took his already-full schedule apart to spend hours studying and experimenting with it. He found that when a person talked or sang into the mouthpiece the vibrations of his voice made tracings on a sheet of glass covered with lampblack. Also one of the professors at the Institute let him borrow something called a "manometric capsule," and back at the Sanders house Aleck demonstrated it to the intrigued Thomas Watson.

"It has an enclosed gas flame, you see, and when you speak into it like this"—he bent his tall body over the capsule— "the vibrations of your voice cause the flame to flicker—so —" He motioned Tom to move nearer and glance over his shoulder to observe the little flame wavering up and down and reflecting itself in a moving, shining surface.

"That's a revolving mirror, Mr. Watson." Aleck indicated the moving ribbon of light. "Now if I can only discover the shape of the elements of English speech I could combine this with the phonautograph and photograph them for my pupils!" He was running his fingers through his heavy curl as he turned excitedly to Thomas. "Do you see? I could use the glass plates of the phonautograph as negatives to depict the proper sounds, and print the tracings for the deaf to study, and then they could speak into the capsule until the flame duplicated the tracing, and then they would know they were uttering the sound as it should be spoken!"

It seemed beautifully simple, and proved to be heart-breakingly impractical. Over and over Aleck would sing the vowel "E" into the capsule and the phonautograph, but it was absolutely impossible to make the capsule's reflections or the phonautograph's tracings uniform. They would vary aggravatingly. It was one of the rare projects he had to abandon, but although he didn't have the comfort of know-ing it at the time, this project was to lead him into some-thing far greater and more breathtaking than any speech-writer could have been.

After Aleck's Canadian Christmas vacation early in 1875, he and the loyal Thomas were toiling late in the Williams shop one night, trying to eliminate a mysterious snag in the multiple telegraph, when Aleck paused, looked thoughtfully at the young man beside him, and then decided to tell him of a startling idea that had been crystallizing in his mind since the summer. "Mr. Watson, I've another idea I haven't told you about that I think will surprise you. . . ."

"Yes, Mr. Bell?" Tom was busy with the troublesome wires, and to tell the truth, he wasn't in a mood to welcome more of Mr. Bell's surprising ideas, but the next words brought him to amazed attention. "I have had an idea in my head for more than a year by which I am sure I will soon be able to talk by telegraph."

Young Watson never forgot the terse, crisp sentences in which Bell was pouring out his inspiration. "Mr. Watson, if I can get a mechanism that will make a current of elec-tricity vary in its intensity as the air varies in density when a sound is passing through it, I can telegraph *any sound, even the sound of speech!*"

The hours slid by, but the two of them forgot the clock

as they talked excitedly. Aleck made hasty sketches of the apparatus as he conceived it, and they planned how they would build it. Aleck had even named it. It would be a "telephone," from two Greek words, *tele* ("afar off"), and *phone* ("sound"). And then came the discouraging question of what it would cost. Sadly and reluctantly they were forced to agree that the wonderful idea must be put aside. "Mr. Sanders and Mr. Hubbard would not welcome such a will-o'-the-wisp idea, and they are pressing me—and rightly so—to complete this telegraph matter," Aleck admitted. "Until I do, I have no right to consider building a telephone."

∽ 9

"Get It!"

SHIVERING FROM THE NEEDLE-SHARP CHILL OF THE FIRST DAY
of March, Aleck negotiated, as neatly as he could, the little
swamps of mud that seemed to make up Washington's
streets, and proceeded to pick his way toward the famous
Smithsonian Institution.

Although it had been the capital of the United States for
seventy-odd years, the city still had a raw, unfinished and
unkempt look. Even the monument that had been designed
to honor the beloved first President stood stubby, neglected
and unfinished because funds had run out, and Aleck understood that Congress had bickered about appropriating more.
He thought of the dignity of Boston and the gracious atmosphere of Cambridge, and looked about him in distaste. As a
residence Washington would not be for him!

Still, it was an important city. He had come to see about
applying for the patents on his telegraphs. Mr. Sanders had
advanced the price of the train fare, Mr. Hubbard had
generously offered the use of his Washington house, and
now that the patent business was under way, Aleck was
striding toward the man he was most eager to see in Washington, Professor Joseph Henry, Secretary of the Smithsonian, the man his friends at the Massachusetts Institute of
Technology considered the most brilliant physicist of the
time.

The Smithsonian itself, built of weathered brown stone

96

and adorned with many turrets, rather reminded Aleck of a small country castle in England or Scotland. He sent in his card and his letter of introduction from one of the Massachusetts professors, and in an amazingly brief time he was standing before Professor Henry himself.

"I am very glad to see you, Mr. Bell," the Secretary was saying cordially. "My friend tells me most interesting things about you. Sit down, please, and tell me how I may help you."

For all the cordiality, Aleck's spirits sagged a little. Looking at Professor Henry he realized that "he was very aged—about eighty years old," as he wrote to his parents, and rather frail. Besides he was suffering from a very heavy cold, and looked utterly forlorn. Aleck wondered if this was a propitious day. But there was no help for it now, so he plunged in.

"I have been working on many experiments in telegraphy, and some of the points that I thought were my own discoveries in acoustics, I learned at the Massachusetts Institute, had previously been discovered by you; so I thought I would explain all my experiments and learn what's new and what is old."

The old man nodded, and then assumed an inscrutable expression as Aleck started to pour out his story. Henry had probably dealt with scores of enthusiastic young scientists before, but suddenly his face was alight with interest and his whole being seemed to leap to attention so abruptly that Aleck was startled. He had just made what seemed to him an unimportant remark. "When I passed an intermittent current of electricity through an empty helix of insulated copper wire, a noise could be heard proceeding from the coil."

"Is that so?" The professor was interrupting him. "Is that so? Will you allow me, Mr. Bell, to repeat your experiments and publish them to the world through the Smithsonian? Of course giving you the credit for the discoveries."

Waves of incredulous delight swept through Aleck Bell,

but he managed to say exactly the right thing: "It would give me extreme pleasure, sir, and I can show you the experiments myself. I have the apparatus with me here in Washington."

"You have?" Professor Henry had forgotten the misery of his cold, and he started to his feet. "Could you show me now? If I should go with you to your lodgings, could you show me now?"

"I do have everything in readiness at Mr. Hubbard's house, yes, sir."

Professor Henry was reaching for his coat. "Then I will go with you immediately! Have you a carriage here?"

Private carriages were not numbered among the Bell modes of travel. Aleck flushed and shook his head, and the kindly professor was quick to understand. "No matter. I will order my own——"

He reached for the bell-rope, but Aleck put out his hand. "It is a raw, damp day, sir, unfit for you to be out. I would rather bring the apparatus here for your inspection at any time you say."

At noon the next day Bell set the instrument working, and for a long time Professor Henry sat at the table holding the empty spiral of wire against his ear, listening to the sound it emitted.

Breathless with the evidence of the famous man's interest, Aleck made a sudden decision. "Sir, I would like to ask your advice about another instrument I have in mind. . . ." Rapidly, and in terse, clear words, he sketched his idea for transmitting the human voice by telegraph while the older man listened with intent eyes in an alert face. At the end Aleck drew a long breath, paused for a second, and then asked, "What would you advise me to do—publish the idea and let others work it out, or attempt to solve the problem myself?"

Henry's voice was quiet, but it had an authoritative note as he answered quickly, "I think you have the germ of a great invention. I advise you to work at it yourself instead of publishing it."

Aleck hesitated, and then confessed the thing that was a very real lion in his path. "I'm very much afraid I do not have the electrical knowledge necessary to overcome the difficulties."

The Secretary looked at him appraisingly, and then his face relaxed and he leaned forward and said two words so emphatically that when he jubilantly reported the interview to his parents, Aleck capitalized and underscored them: *"GET IT!"*

That was just one day before Aleck's twenty-eighth birthday, March 2, 1875, and he knew then that he would—he must—work out his idea for an electric, speaking telephone regardless of what his backers might say or think, and despite all the other obstacles in his way.

He returned to Boston and Salem elated, determined—and very poor. It was true that Mr. Hubbard and Mr. Sanders were paying for the costs of his experiments and patents, but they were not paying him for his time, and as he plunged deeper and deeper into his scientific investigations he had less and less time to give to his teaching. By the spring of 1875 his classes were, as he put it himself, "utterly disorganized."

Just at this time the only assets Aleck had were his friends, but they make an impressive roll call.

There was Sarah Fuller, who kept his accounts and introduced him to a noted ear surgeon, who could give him important aid in his research. There was Mrs. Sanders, a motherly soul who worried about his scanty meals and the too-few hours of sleep he permitted himself. Perhaps she also made pessimistic remarks about his inventions, because Aleck once cried out, "I live too much in an atmosphere of discouragement . . . Good Mrs. Sanders is too much in the habit of looking on the dark side!"

Nevertheless it was Mrs. Sanders who surreptitiously cut his candles short so they would burn out sooner and force him to go to bed, and who carried trayfuls of nourishing food up to the third floor and slipped them unobtrusively inside Aleck's door.

There was Thomas Sanders, who not only gave his financial backing to the Bell ideas, but, when he happened to be visiting at his mother's home, submitted to being waked in the middle of the night to listen to some hoped-for improvement in the telegraph.

"Mr. Bell would wake me with his dark eyes blazing," Sanders said a little wryly, "begging me to go down to the cellar, and then he would rush wildly out to the barn and begin to send me signals over his wires. If I noticed any improvement . . . he would be delighted, and would leap and whirl in one of his war-dances, and would go contentedly to bed, but if the experiment was a failure he'd go back to his work-bench."

Perhaps Aleck's greatest asset was Thomas Watson, with his loyalty, his quick understanding of the designs Aleck wanted made, his deft sureness in producing them and his supply of humor. Most important was the enthusiastic interest that kept him working beside Aleck Bell in the gloomy, dusty attic of the Williams Shop long after his lengthy day in the shop itself was over. "We were there sometimes eighteen hours at a time," Tom Watson confessed afterward, "at least, I was!"

The multiple telegraph was not an easy invention to bring into the world. By the spring of Aleck's visit to Washington both men had lost faith in it. "The apparatus sometimes seemed to be possessed by something supernatural," Watson commented later, "but I never thought the supernatural being was strictly angelic!"

Apparently the thought of washing his hands of A. Graham Bell and his projects never crossed Watson's mind, for which the entire world has reason to be grateful.

It was exactly three months after Professor Henry's laconic "GET IT!" The two men had been working in the dusty, sweltering attic through the late hours of the early

June afternoon—Aleck in one room, with the crude receiver pressed to his ear; Tom in the far end of the other, sending him signals. For some reason one of the transmitter reeds stopped vibrating and Tom plucked it to set it going. It had been screwed too tightly, and he was still plucking at it to set it free when Aleck burst in upon him "with a great shout."

"What did you do? What happened? Show me!"

Tom obeyed, and then looked at him in astonishment. What did it matter that the one reed had stopped?

It mattered a great deal. The wires had not carried the intermittent current, as Tom himself described the situation later, because Tom's reed had frozen, but the little strip of magnetized steel which he had been plucking was generating by its vibration over the electro-magnet "a sound-shaped current," and Aleck's musically-trained ear, made acute by his long experiments with the telegraph, had caught the faint *twang* that would have eluded anyone else.

Aleck knew then that he had the key to his telephone—not the make-and-break intermittent current he'd been working with, but one that would carry sound by a continuous undulatory current!

Carried away by enthusiasm, the two men spent the rest of the afternoon and early evening repeating the test with every kind of transmitter they had. And it worked! Satisfied at last, they forgot the stifling heat in the attic and their tiredness, and fell to making plans to construct the first telephone. Their tests had been with telegraph instruments, but now they needed a mouthpiece and a receiver. They talked far into the night, and Bell made sketches of the thing he wanted.

"Could you have that ready by tomorrow night?"

"Of course!" said the confident Watson.

They stumbled into the midnight train to Salem, but neither of them slept that night. Aleck wrote an exultant letter to Mr. Hubbard, and Tom was too busy constructing his telephone to sleep. He was back at the shop as early as the first train would take him to Boston, and made every bit of that precious first instrument with his own hands. When Aleck saw it he announced it was perfect.

Their attic rooms over the shop wouldn't offer a fair test—they would be too close together—so the resourceful Tom strung a telegraph wire from the attic to his bench, two floors below. After the shop was closed they made the great experiment, but not a sound could they wring out of the telephone.

Still Aleck knew they were on the right track. Something still eluded him, but he had an inner sureness about this telephone he had never felt about anything else. This seemed to be the beginning of a propitious summer—and then came the deluge.

Tom Watson fell ill of typhoid (both men had been driving themselves at such a terrible pace that it is something of a wonder either of them survived), and in consequence all work on any of the inventions stopped. Aleck himself was on the verge of nervous exhaustion, and to make matters worse the pressure of his debts began to close in on him. At this point his only income was his very slender salary from Boston University, but he was too fiercely proud to confess his need to either Thomas Sanders or Gardiner Hubbard, who would certainly have come to his aid. The simple Biblical-sounding phrase he used later to describe his plight tells its own stark story: "I began to be in want."

In desperation Aleck did appeal to the head of the School of Oratory of Boston University, and to Professor Munroe's everlasting credit he advanced the inventor the fees for the next year's lectures, a tribute to A. Graham Bell's very evident integrity. It was an act Aleck never forgot. "Without this aid," he said later, "I wouldn't have been able to go on at all."

But then he learned the thing that actually did make the world cave in around him—Mabel Hubbard was leaving Cambridge to spend an uncertain number of months with Cousin Mary Blatchford.

The Sunday hours Aleck had spent with Mabel had become oases in his life. He had come to depend upon her blitheness; her intelligent interest in what he was doing—or trying to do; her happy gift of understanding. She revived him as no one else did.

He had kept a careful guard against betraying the secret of his love and hopes to Mabel or anyone else in the two years since Mabel had first come to the little basement classroom. Indeed, only a few weeks before, he had been writing to his parents out of the depths of a despondent heart, "I doubt if I am the marrying kind of man." But the shock of the news of Mabel's leaving Cambridge had taught him differently, and it was no longer any use to try to ignore or subdue the force that had been building up within him these past two years. Mabel was going away. In his misery he wondered if she was saying good-by to Cambridge forever. He had to tell the Hubbards; he couldn't let Mabel go without having her know how dearly she was loved.

A more prudent and realistic man would have stopped to consider that when all his projects were in a very questionable condition and he himself was not only half ill but burdened with debts, it was hardly a propitious time to approach parents and ask for a daughter's hand. But Aleck Bell was feeling neither prudent nor reasonable. He took the first possible street car to Cambridge to call on Mr. Hubbard in the Brattle Street library, and felt that the fates were against him when he discovered Mr. Hubbard was out of town. The words froze in his throat when he thought of saying what he had come to say to Mrs. Hubbard alone. But when he had beaten a retreat to the Sanders' home, he managed to take a tight grip on his courage and poured out his desperation in writing:

June 24, 1875

Dear Mrs. Hubbard:

Pardon me for the liberty I take in addressing you at this time, but I am in deep trouble, and can only go to you for advice.

I have discovered that my interest in my dear pupil—

Mabel—has ripened into a far deeper feeling than that of mere friendship. In fact I have learned to love her very sincerely.

It is my desire to let her know now—how dear she is to me, and to ascertain from her own lips what her feeling toward me may be. Of course I cannot tell what favor I may meet with in her eyes. But this I do know—that if devotion on my part can make her life any happier—I am willing to give my whole heart to her.

I called yesterday for the purpose of consulting the wishes of Mr. Hubbard and yourself. Mr. Hubbard's unexpected absence rendered me undecided what to do.

The date of Mabel's departure is so near, and her father's return so uncertain that I think it best to address this note to you.

I am willing to be guided entirely by your advice. . . . I promise aforehand to abide by your decision however hard it may be for me to do so.

> *Believe me, dear Mrs. Hubbard,*
> *Yours very respectfully,*
> *A. Graham Bell*

✌ 10

Tempestuous Summer

FOR TWELVE YEARS GERTRUDE HUBBARD HAD BEEN STRIVING to help Mabel enter into the world around her and be ready to take her part in any advantage or experience that would come to a normal girl, but now that the greatest experience of all was offered to her, it was Mabel's mother who felt shaken and very unready.

It had crossed her mind that young Gertrude—Sister— and Mr. Bell might be attracted to each other someday, but Mabel—she knew her husband had said sadly he thought it highly unlikely that May would ever marry, and after all, she was much too young for anyone to consider it. . . .

She looked across at Aleck, who had come in response to her summons, and suddenly saw him with understanding eyes.

"Mr. Bell," she spoke very gently, "you have caused me the greatest surprise any man could have given me, but I may say I am not displeased. We have known you intimately for well over a year, and I have never seen nor heard anything to alter my opinion that you are a truly fine man. I do think, however, that Mabel is not ready—not old enough —to hear your confession. And I feel sure her father will be of the same opinion. If you spoke now I fear it would startle her, and being unprepared she might decide in a childish way, without fully knowing her own mind. So I ask you to wait a year, and if your mind is the same at the end of the year, you may speak, and meanwhile I promise that you will be given full opportunities to see her."

She could fairly watch the tautness and anxiety ebbing out of the tense figure before her, but noting the drawn lines of Mr. Bell's face she felt impelled to raise a delicate question.

"You know, Mr. Bell"—Oh, how could she phrase it most tactfully?—"we would want her to enter society—meet other people—younger men——"

Aleck flushed unhappily. "You probably take me for considerably older than I am, Mrs. Hubbard. I am twenty-eight."

And now it was Mrs. Hubbard's face which lighted with relief. "I am so glad," she said joyously. "You must forgive us—but none of us thought you less than thirty-six! Mr. Bell, it is Harvard Class Day, and of course you know Mr. Hubbard is away. Would you be willing to escort us Hubbard womenfolk to the celebration?"

Would he? He sprang up, weariness and doubts and debts forgotten, and whether it was Mrs. Hubbard's good offices or not, he seemed to have Mabel all to himself at the Class Day celebration. "She evidently took some pleasure in my society," he wrote Melville and Eliza Bell, "but was perfectly unconscious of what was passing in my mind!"

Two days later, on a Sunday, Gardiner Hubbard returned, and was even more astounded than his wife had been. Apparently what had been so evident to Berta's and Grace's sharp eyes months before had eluded Mabel's mother and father completely.

Mr. Hubbard promptly began to act like a father. "I consider Mabel such a child that if Mrs. Hubbard hadn't said 'one year,' I would have said 'two'!" he told Aleck.

Then he studied the young man gravely. He had dealt with all varieties of men, had Mr. Hubbard, in Cambridge, Boston, Washington and abroad, and he recognized a trustworthy man—and a gentleman—when he saw one.

106

Aleck Bell was too thin, too intense, too impetuous and possibly too visionary, but Mr. Hubbard liked what he saw and what he knew. And if Mr. Bell said, as he did say in his note, that he was willing to give Mabel his whole heart for a lifetime, it would be a good gift, clean and staunch and enduring. The appraising light went out of his eyes and he clasped Aleck's hand warmly.

"Now I will tell you, Mr. Bell, I've had quite an affection for you from the very first time I met you, and it was this that led me to offer my assistance in regard to your telegraph. I believe you have very great talents, and my object in aiding your telegraph schemes was not a speculation but the hope of encouraging you to devote yourself to science." He paused, and then spoke emphatically and earnestly. "Mr. Bell, I believe you have a great future before you. But," and his eyes twinkled for a moment, "that does not mean I wish ideas of marriage or engagement put into Mabel's head! Not at the present time, at least, and of course you need not consider yourself bound in any way."

His wife had come to sit beside him, and he took her hand in the gesture familiar to everyone who knew them. "Now I think we are finished, Mr. Bell—unless you have something to add, my dear?"

Mrs. Hubbard smiled at him. "Only to tell Mr. Bell that I heard Berta and Lina talking about going into the garden, and May said she would like to wait for Mr. Bell!"

Aleck sprang up with alacrity and hurriedly excused himself. He was possessed by two emotions as he joined the girls, and he and May followed Berta and Lina McCurdy into the rose-scented garden: it was an inexpressible relief and comfort to know that the Hubbards really thought well of him, but to be bound by a promise not to mention what was uppermost in his heart for an entire year—already he was beginning to rebel!

He was elated to be with May in this enchanted place of flowers and night scents and to feel the closeness of the girl clinging to his hand to steady her in the dusky light,

107

yet he felt ashamed and embarrassed at the thought he was trying to win her without her knowledge, and he resented the promise he had given to Mrs. Hubbard.

"I was happy to have you near me," he told her later when he wrote a full confession of his behavior, "happy to feel your hand upon my arm—and above all happy and thankful that I could talk to you, and that you could understand me although the moon afforded such a feeble light, but it was *so* hard not to be able to tell you. I felt it was unjust to you—that it was *wrong* that I should be with you so without your knowledge and consent."

The others had skipped lightly ahead, but now they were darting back, each holding a single, long-stemmed daisy. "Try your fortunes!" Lina challenged. "Think of the ones you love and see if they are true!"

Mabel accepted hers with innocent readiness, and with Berta and Lina standing by with such mischievous looks Aleck did not dare refuse. "You first!" they prompted him, "Now—she loves me—she loves me not—she loves me——"

When his reluctant plucking indicated that all was well with Aleck's true love, the two girls glanced at each other and giggled, although May's eyes were still untouched by any suspicion as she asked, "Who is she, Mr. Bell? Do we know her? Do tell us her name."

Aleck had given a promise, but he was tempted almost beyond his strength. The face he turned toward Mabel in the moonlight was grim. "No!" the word came explosively, "No—no, Miss Mabel—I can't tell you!"

Berta and Lina burst into gleeful laughter as they looked at May. "We know—we know!" they chanted and ran off toward the house leaving a very embarrassed couple in the midst of the garden. Somehow both the garden and the moonlight had lost their magic.

"How different was this from what I intended!" Aleck was to exclaim in his letter. He felt shamed and humiliated, as though someone had intruded upon a very intimate thing before it was clothed and ready to be seen, and looking at the change in his face Mabel was vaguely disturbed and uncomfortable.

"What is it, Mr. Bell? What has happened? Is it one of your headaches?"

Aleck could only shake his head in misery. "No, no— my head is very well. Forgive me. I can't explain. Shall we go back?"

His misery followed him on the verandah, where he sat at Mabel's feet so that the light from the house might shine on his face. Everyone in the Hubbard family had been drawn out to the verandah's pleasant aspect, and Aleck's flesh fairly crawled at the knowledge that they all knew his secret—all of them except the one person he felt had the right to know. Added to that, he was beginning to have the unhappy conviction that there was very little if anything about him that could possibly appeal to this gently-bred, cultured and lighthearted girl of seventeen, anyway. If only he could know what her ideal was, perhaps, in his year of silence, he could work toward making himself like it. And then the thing happened that turned the rest of the summer into a nightmare for him, for to his great horror he heard his thoughts burst out into words. "Miss Mabel— if you could choose a husband, what would you wish him to be like?"

He heard the gasp from all parts of the porch and saw the quick widening of her eyes, and he plummeted to the depths of humiliation, although Mabel came to his rescue with some playful remark, and even Aleck forced himself to join in the answering laughter. But he had done an unpardonable thing. He had embarrassed Mabel before everyone, probably offended her father and mother, seemingly broken his promise and certainly behaved with unspeakable rudeness.

There was no sleep for Aleck Bell, nor rest, for many days. Perhaps, had Mabel remained in Cambridge, the unhappy business would have straightened out within a few days; but on Friday when she left to spend the summer in

Nantucket with Cousin Mary and Aleck was alone, he was at too low an ebb even to plan to go to his parents for his usual vacation.

Instead he wrote them a full story of his long-growing love for the girl who had come to him as a pupil, and his interviews with the Hubbards, but the Sunday night episode was too sore a subject to mention. At the end he begged, "Dear Papa, please give me your sympathy—I love you very much."

For some reason Melville Bell did not answer, and Eliza Bell, so her son said, "barely alluded to the matter, and spoke of Mabel as if I had called her a deaf-mute." Aleck felt as though there was no joy left in the world.

Perhaps he was exaggerating the enormity of his offense out of its true proportion, although in that mid-Victorian era young men did not ask such questions as Aleck's lightly. This, together with the fact that he bitterly regretted binding himself with a promise, gave Aleck a very bad summer. He had too high and forthright a code of honesty to feel it was honorable to continue seeing Mabel and try to win her heart without laying all his cards upon the table. He stood the misery for a month, and then his diary, in a few staccato sentences, gives the results of another visit to the Hubbards:

Friday, July 23. Called on Mr. & Mrs. Hubbard. Told them I could not help myself. Should tell Mabel what was in my mind if they gave me an opportunity. Thought she should know it. I could only avoid telling her by avoiding her. Should tell her the first chance I got. Would go to Nantucket for the purpose unless they forbade me. They said:—Feared Mabel would be startled and distressed. Wanted her to be near her mother.—Would be home August sixth. If I would wait—would give me an opportunity to speak to her. Agreed to wait.

Thursday, July 29th. Private conversation with Mrs. Hubbard in which she stated that Mabel had taken a

dislike to me. Did not like long hair. Preferred blue eyes!
Don't know what to do. Quite distressed about it—such
dislikes have a tendency to be permanent!

The young man was too desperate to remember from
his Shakespeare that it is a good sign when the "lady doth
protest too much."

In Nantucket Mabel herself was having her own un-
merry time. Cousin Mary Blatchford had seen certain signs
in Cambridge that showed her only too well which way
the wind was blowing for her adored young relative. Cousin
Mary was a soul who scorned diplomacy, and one day she
flung a bombshell question at Mabel. "May, are you willing
to be the wife of Mr. Bell if he asks you?—and he will ask
you! I have been very sure for some time that he wants
you!"

The startled girl dropped her sewing and could only
color slowly. "I—I don't know, Cousin Mary"—she was
floundering for a safe answer—"Really, I haven't—a girl
shouldn't think——"

Cousin Mary smiled grimly. "Then I advise you to begin
thinking! I consider it your duty to make up your mind
here so as to be ready when you go home. Especially if you
consider Mr. Bell an honorable man."

She changed the subject then, but Mabel's sewing did not
go well the rest of the afternoon, and her cheeks were hot.
They didn't mention the matter again, but it wasn't easy
to forget anything Cousin Mary meant one to remember.
A few days later Mabel, like Aleck before her, was sum-
moning her courage to write a letter to her mother. It was
a difficult letter, and the words didn't come easily. There
were many discarded sheets. She began by repeating Cousin
Mary's remarks, and then plunged on:

I think I am old enough now to have the right to know
if he spoke to you or Papa. I am not much of a woman
yet, but I feel very, very much that this is what it is to

111

*have my whole future life in my hands . . . I think I am
very, very impressed by the honor done me. It is so strange
that anyone should consider me fit to be his wife.*

It was a fine, brave and thoughtful letter for a girl of
seventeen, and in its few pages Mabel Hubbard grew from
a girl into a young woman. And then came a final, childish
protest:

*Oh, Mamma, it comes to me more and more that I am
a woman such as I did not know I was. I cannot get over
the surprise and feel more and more how unfitted and
unworthy I am for such a position as a woman and wife.*

*Of course it cannot be, however clever and smart Mr.
Bell may be, and however honored I should be to be his
wife, I never, never could love him or even like him
thoroughly. . . . I wish I had you here that you might
speak to me sensibly and clearly and tell me how I ought
to feel . . . I must hear from you to clear me up, I feel so
misty and befogged. . . . Help me, please.*

Your own Mabel

Back in Cambridge Mrs. Hubbard summoned Aleck and
read him part of the letter. The rest, she told him, would
be painful for him to hear. What really was painful was the
news that Cousin Mary Blatchford had changed her mind
about bringing Mabel home on August sixth, and was keeping her in Nantucket for another fortnight. That did it.
Aleck spent a restless night, snatched a few things together,
caught the six-thirty street car to Cambridge and re-
appeared before a startled Mrs. Hubbard, valise in hand.

"Why, my dear Mr. Bell, what is it? You are leaving us?
Are you on your way to Canada? Come in; you must take
breakfast with us before you go. Is this a sudden decision?"

Aleck shook his head wearily. "It is a sudden decision,
but I am not on my way to Canada. Oh, Mrs. Hubbard, I
cannot bear what you told me yesterday—that Mabel dis-

Gardiner Greene Hubbard, staunch advocate of oral education for the deaf, and sponsor of the Bell telegraph and telephone.

Painting by W. A. Rogers

Alexander Graham Bell and Thomas Watson in the wire-strung labora-tory at 5 Exeter Place, where the first telephone message was transmitted.

Alexander Graham Bell at work in his Washington study, soon to be the scene of the disastrous fire.

Woodcut from the Scientific American, March 31, 1877, showing "Prof. Graham Bell" before an audience at Salem, Mass., telephoning to Thomas Watson in Boston, 14 miles away!

A duplicate of Alexander Graham Bell's first telephone.

The Bell family in 1885. Elsie, at the left, become Mrs. Gilbert Gros-venor; Daisy, Mrs. David Fairchild.

trusts me. I am bound for Nantucket. I must go. And I have come to ask you to accompany me."

He looked so ill, so distressed and forlorn, that Mrs. Hubbard laid a gentle hand on his arm and drew him inside the room. Her face was very kind. "I recommend that you wait patiently for Mabel's return. In any case it is impossible for me to go with you, and in your present state I think it would be a great mistake for you to try to see Mabel. I think you are truly ill, and should go to your friends in Canada until you are recovered—for all our sakes."

All the Scotch stubbornness rose in Aleck. He shook his head obstinately. "I cannot go to Canada until this mistrust of Mabel's is cleared. I must go. If she doesn't wish to see me I shall return. I don't want to go in an underhand way, but openly and honorably. If you wish, you may write her telling her not to receive me. Perhaps it would be well for you to write her a few lines in case she doesn't care to see me?"

Here was an overtense young man whose impetuous impulses and passionate torrent of words were enough to give any mother pause, but suddenly Mrs. Hubbard was catching a glimpse of a soul so strong and finely tempered that its possesser would deliberately risk losing what he craved most in the world rather than gain it by what he considered an underhand method.

She answered him gently. "Mr. Bell, I think you are entirely right to do what you think best. I happen to think you are mistaken, but if you feel you are right, you must go. I think there's no need of a note to Mabel, but I will write a few lines to Miss Blatchford so that there will be no difficulty with her if Mabel wishes to see you."

So Aleck went to Nantucket. He found that Siasconset, where Mabel was, was eight miles away, and "Captain Baxter," who drove passengers between the two towns, didn't make the trip on Sundays. Besides, a storm came up,

113

"the heaviest fall of rain I ever saw." Very well, he would use Sunday evening to write Mabel a letter!

Its length was far beyond the limits of the usual letter, but it wasn't a usual letter. In it Alexander Graham Bell opened his whole heart and soul to Mabel Gardiner Hubbard. He confessed the whole story—his first letter to her mother; the promise he had given her; the full meaning of the unhappy Sunday evening episode that had cost him such misery all summer. He told her that he wanted to be as free and open with her as he had been with her parents —that he felt it was wrong for him to visit her disguised as a friend of her father's. He told her that he loved her with "a passionate attachment that you cannot understand, and that is to myself—new—and incomprehensible." He wanted her to be his wife and share his life, but he did not ask her to marry him, or even to say that she loved him. He wouldn't ask either question until she had an opportunity to know and understand him better. The only thing he did want was permission to try to win her affection if he could. He wouldn't even mention the other matters until she was ready.

But about her distrust of him—oh, that was something else again. He couldn't bear to have that shadow between them. Wouldn't she see him and help clear it away? He ended the letter with "I await your answer as to whether I may see you or not. There is only one question I want you to answer. I would like you to tell me frankly all there is in me that you dislike and that I can alter."

Mabel did not see him: Cousin Mary Blatchford saw to that. When he presented himself at the Blatchford cottage a day or so later he was met—greeted would hardly be the word—by a very starched Cousin Mary. She wrote about the encounter to a friend, "One day he presented himself and I took fright and dismissed him." Then she rustled upstairs "and talked it over with Mabel, and more or less told her that she *did not* care for him!"

The tide finally turned, however, when Mrs. Hubbard sent Aleck a very brief note dated August 20:

Dear Mr. Bell:
   Mabel came home last night most unexpectedly . . . I cannot tell you how nearly we lost our darling in the undertow on Wednesday. Shall we see you as usual on Sunday p.m.?

Would they see him? They would indeed! And when he strode up the steps and appeared before the door, and the door swung open before he could lift the knocker to reveal a blushing, trembling Mabel on the threshold, he felt as though the door of Heaven itself had opened. He could only stand gazing in incredulous delight.

"Mr. Bell"—her words were more blurred than usual because of the faltering of her voice, and she had to stop to try to control it—"Mr. Bell—would—would—you come into the greenhouse with me—where we—can be—alone?" she turned away before he could answer, which was perhaps as well, because Aleck was trembling violently himself, and it would have been difficult just then to control his lips. Safely out of earshot, and having given a quick glance around to be sure they were not seen, the girl lifted her face bravely to the eager eyes above her.

"Mr. Bell—" Oh, why, when she so wanted to speak clearly, did her throat have to close? "I—I really don't dislike you! I—I like you! I don't love you—but I do like you! Is that enough—now?"

It was a rather prim and stilted return for his outpouring, but the sight and the nearness of the girl restored Aleck's usual sanity. He had the sudden insight to know what had happened. "She has been for two months under the care of a guardian who had a dislike of me. An *old maid,* who has no thoughts of—or sympathy with—love. She had been so influencing Mabel against me that she had formed a deep dislike against me and distrusted my intentions."

115

But now they had a "full and free talk in the greenhouse," and all the clouds that had been hanging over them in the summer were dispelled. They would gladly have remained in the safe retreat of the greenhouse, but eventually they had to assume noncommittal expressions and face the circle of smiling, knowing faces around the tea table. Aleck commented rather wryly in his diary, "Great trial to be among so many ladies with the knowledge that all were conscious of the object of my visit. Quite proud of Mabel for the way she stood the trial."

He also recorded how close she had come to being swept away from those who loved her. "She was thrown down in the surf, dragged out by the undertow and rescued in a semiconscious condition." He was horror-struck by the thought. "Had she died I would have committed suicide," he added grimly.

He doesn't mention it, but a few days later another note from Mrs. Hubbard must have given him cause for one of his famous war dances: "I give you back your promise entirely, unreservedly . . . If you can win Mabel's love I shall be happy in my darling's happiness."

It was then that he thought he could safely return to Canada for a few weeks' respite from the new season's rounds of lectures and working on inventions.

When he came back, Mabel was so undeniably happy to see him that he thought nothing could ever dampen his days again, but he discovered very speedily that he was wrong. For once, and for possibly the only important time in their lives, Mr. Hubbard's understanding and shrewd foresight failed him. He was disappointed and dismayed at the way Aleck pursued that will-o'-the-wisp, the telephone. Perhaps he couldn't be blamed, for it had been June when Tom Watson built the first telephone, and it was now November—and the thing hadn't, as Aleck ruefully admitted, transmitted even a nearby sound, let alone lived up to its name and carried a faraway one.

At any rate Mr. Hubbard said bluntly and forcefully that

Aleck and Thomas Watson were wasting altogether too much time and money on the contraption they called a telephone, which certainly would never get either of them anywhere. Furthermore, Mr. Hubbard thought it was high time Aleck abandoned his Visible Speech classes. He should settle down to one thing, and that was the perfection of his telegraph ideas, and then devote himself to the further study of science. He really had a future there.

If the truth were told, Aleck himself was tiring of teaching and lecturing. The thought of devoting himself to science was very attractive, but to give up the telephone— his most ambitious invention, the most challenging idea he'd ever had? Not A. Graham Bell!

Undoubtedly Mr. Hubbard thought he was acting in everyone's best interests, although it is probable that he was thinking most of all as Mabel's father, who flinched at the thought of her marrying an inventor whose ideas turned out to be mirages all his life, so he issued a parental declaration. Mr. Bell could take his choice: become a sensible man, admit his telephone scheme was sheer fantasy, turn his Visible Speech classes over to someone else, settle down to the telegraph projects in earnest and marry Mabel; or cling to the classes, go on with the telephone—and give up Mabel!

Aleck Bell hadn't the slightest intention of doing either. He answered Mr. Hubbard with a unique combination of polite respect and flaming determination:

> *I shall certainly not relinquish my profession until I find something more profitable (which will be difficult), nor until I have qualified others to work in the same field. I confess there are many things about Visible Speech that I regret, and that I would change were I free to do so, among other things the name itself, and I confess that the sort of labor involved in teaching is too mechanical to be much to my taste. I should infinitely prefer being employed in other ways.*
>
> *Should Mabel come to love me as devotedly as I love*

117

her—she will not object to any work in which I may be engaged as long as it is honorable and profitable. If she does not come to love me well enough to accept me whatever my profession or business may be—I do not want her at all! I do not want a half-love, nor do I want her to marry my profession!

He paused, a little distressed at the vehemence of what he was writing, yet still convinced that in this instance Mr. Hubbard was entirely wrong. He began again, with the winning frankness it was so hard for anyone who knew him to resist:

I suppose I must have said a great many things I did not intend to say when we talked. I shall feel sorry if I have given you the impression that I do not respect your opinions and feelings, for I do, very sincerely.

And then he touched upon a matter which his acutely sensitive nature must have needed to make very clear:

You are Mabel's father—and I will not urge you to give —nor will I accept it if it is offered, any pecuniary assistance other than that we agreed upon before my affection for Mabel was known.
I appreciate fully your feelings in regard to my actions, and I am sorry I am compelled by circumstances to act as I do. Please bear with me a little longer?

There the matter rested between the two men, but it had a surprising result. Mabel had not enjoyed the thought of being weighed against Aleck's telephone. The day after Aleck dispatched his letter was Mabel's eighteenth birthday and also Thanksgiving Day, a fact Aleck was to record joyously in his diary. He came to Cambridge to join in the double celebration, and once again Mabel was waiting for him. She drew him inside, out of the New England November chill, and once again guided him to their favorite spot in the greenhouse. What happened there she confided to Mary True in a letter:

I told him that I loved him better than anybody but Mamma and if he was satisfied with so much love I would be engaged to him that very day! [Was this the same girl who had been talking about preferring blue eyes, and protesting that she could "never love Mr. Bell, nor even like him thoroughly" only a brief three months ago?] *I took him by surprise, he said I had come to him at the very lowest point of despair, I had seemed so far away from him with so many things between! He almost refused to let me bind myself to him, he reminded me how young I was and how I had not seen other men. But I told him I could never find anyone to love as well, so he submitted very cheerfully to the engagement, only he wants me to understand that it was I who did it, and of course he could not refuse a lady! After he was gone I was frightened.*

The next two days were forty-eight hours of such utter misery for the girl that Gertrude Hubbard was frightened, and wanted Mabel to break the engagement on the spot and escape to New York, but Mabel couldn't do either thing. On Sunday Aleck came again, and Mabel's letter ended on with a note of triumph:

Just as soon as I saw him and had his arms around me I forgot about being miserable, and was quite satisfied that I did love him very much indeed! Since then I have seen him every day, and have been loving him better and growing happier all the time!

The long, tempestuous summer was over at last.

✆ 11

"Mr. Watson, I Want You!"

LONG AFTER EVERYONE IN THE SANDERS HOUSE HAD GONE TO bed on that memorable Thanksgiving night, Aleck Bell sat in his third story room. For once he had no thought of telegraph or telephone. He was thinking of the wonderful thing that had happened that afternoon, and glorying in the fact that Mabel had come to him of her own accord and "proposed" to him despite her father's dictum. Finally he drew a sheet of paper toward him, moved the candle nearer and took up his pen.

He had written Mabel many letters, but except for the one he had written her when they were both in Nantucket not one could be called a love letter. Even the one he had written on the twenty-fourth of November had begun, "My dear Miss Mabel," and was signed decorously, "Very sincerely yours, A. Graham Bell."

This one was different; he didn't need to be discreet or decorous any longer. So he began, "My own little girl,"

> *You must not scold me this once if I do sit up for a moment—even if it is late—to write a few lines to you.*
> *I little thought when I went to Cambridge this afternoon—of the surprise in store for me. You seemed to be drifting away from me—so far away—with Visible Speech and ever so many things between us—and I almost despaired of ever reaching you.*
> *I little thought how near you were. I can scarcely believe now that you really and truly love me—and that you will be my wife. I am afraid to go to sleep lest I should*

120

find it all a dream, so I shall lie awake and think of you.
It is so cold and selfish living all for one's self. A man
is only half a man who has no-one to love and cherish. It
will be my pride and delight—Mabel—to protect you and
love you. Don't go away from me any more.
May God guide us both so that we may be a comfort
and support to each other.

Yours and yours only,
Alec

It was a prayer that was to be more than fulfilled all
the rest of their lives.

Sometime in 1875 Mabel had "entered Society." Her
mother described it in a letter which she unfortunately
dated only "Tuesday, 1875," but in it she refers to "Mr.
Bell." After Mabel's eighteenth birthday he became "Alec"
to the Hubbard family, having dropped the "k" somewhere
along the way in the past few months.

The Hubbard house must have bloomed with flowers
and an air of festivity that evening, although in the afternoon Mabel's courage had begun to quail, and when someone brought a note saying that her dearest friend, Edith
Longfellow, and her sister could not come, she was utterly
forlorn and snatched at the opening to beg, "Let's put the
whole thing off!"

"Everything is arranged, the musicians are engaged, and
everyone is expecting to come," Mrs. Hubbard reminded
her. "I know what Edith's absence will mean, May, but you
must make the best of it!"

So at eight o'clock a rather uncertain Mabel stood in her
proper place with all her usual grace and charm. If she was
nervous it only heightened the glow in her naturally exquisite coloring and deepened the blueness of her eyes.
"She wore a beach silk and full tarlatan plaitings, and
looked her loveliest," said her proud mother, "and received
and introduced everyone with the greatest ease and self-
possession, and I looked on amazed and delighted!"

When the guests had all been presented, May and Harcourt Amory led off in a waltz. "And then Mr. Bell appeared. . . . One look at May's face told how happy she was. I wish you could have seen her, so bright, so fresh, so full of enjoyment and so pretty. . . . They danced the Lanciers, then waltzs and gallops, and I grew young again, and the old house seemed like old times, full of young, bright, merry faces. We had supper at ten, oysters and ices. A few went soon after, but it was past twelve before the last left . . . had the Longfellows come, Mabel's happiness would have been complete."

As her mother was dipping her pen again, Mabel came into the room. Her mother dropped the pen and held out her arms. "Oh, my dear, I cannot tell you how pleased I was with you tonight!"

The girl dropped the apple she had been eating and flung her arms about her mother's neck, half laughing, half crying. "Oh, Mamma, I am so glad! I would rather please you than anything else in the world! But I really felt very queer, and my lips trembled so, and my voice felt so odd I could hardly speak at first!"

"She did not show it," Gertrude Hubbard concluded her account, "and I was amazed."

As the last weeks of the old year ebbed away and 1876 dawned—the year of incredible happenings for Alec Bell—he was back in his old treadmill of teaching the deaf, accepting pupils with defective speech, training teachers in the use of Visible Speech—and working every spare minute far into the night on his telephone.

He had left his comfortable and pleasant quarters at Mrs. Sanders'. He was frankly afraid that uninvited people would invade the Williams attic and steal or copy his instruments and apparatus, so he had moved into two rooms of an attic at 5 Exeter Place, in Boston, and he and Tom Watson fitted up a laboratory there. The place became so much like home to Tom that eventually he quit his parents' home in Salem and took another of the attic rooms in Exeter Place.

By now there was a frequent and very interested visitor to

the laboratory. Young Willie Hubbard, who used to help
Berta and Grace tease Professor Bell, had now entered
Harvard, and was often at his uncle's house. He realized
now that this professor his cousin May was in love with was
the most remarkable man he had ever met. Alec liked him,
and as Willie himself said, "a strong friendship began for
me, and, though I little foresaw it, a part in the greatest advance in electrical science the world has known." Alec invited him to come and see the devices in his new laboratory.

"Alec is working on instruments which will send six or
more telegrams a minute," May informed him.

"I think I can also send music over the wires," Alec struck
in eagerly. And I am very hopeful of an instrument I am
now working on that will make it possible for people to talk
between distant points, such as Boston and Cambridge—
or perhaps even greater distances!"

The music idea interested Willie—somehow that seemed
as though it might be possible—but to carry the human
voice along a wire—He hoped his skepticism wasn't showing too plainly! But he did accept Alec's invitation to come
and see the instruments, and when he had seen them demonstrated was so enthralled that after an evening at the Hubbard house he would often go back to Boston with Alec.
The two of them would have a late supper at the Parker
House, and then go on to Alec's attic apartment to work
over the telegraph and telephone instruments until early
morning. Alec's night-owl habits were as contagious as his
enthusiasm. While Willie hadn't Tom Watson's professional
knowledge and experience and single-minded devotion, he
apparently was mechanically-minded and "neat-handed"
enough to become a valuable assistant. He said later: "These
hours of work together were full of pleasure and companionship; full of the intense interest in the great problem, and
full of the satisfaction of accomplishment."

The attic laboratory was an exciting place, with the multiple telegraph, the autograph telegraph, an apparatus to
transmit music over wires and the telephone scattered all
about, although Willie Hubbard said that Alec "abandoned

the transmission of music as mere play"—if he had persevered he might have anticipated Lee DeForest's radio by about forty years—"and gave every thought to the transmission of voice."

Mrs. Hubbard and Mabel sometimes came to the laboratory. Thomas Watson said he had never seen a man so thoroughly in love as Bell was, and Tom discovered for himself that May Hubbard was a delightful girl.

Now Mr. Hubbard made a surprise announcement. He still didn't believe in the telephone, and even if Alec did prove that the thing would carry speech, he couldn't see any future for it—but he had decided that Alec should take a patent out on it. It was certainly new, and—well, one never could tell. Something might happen, and as a businessman Mr. Hubbard knew Alec's rights to an invention should be protected. He told him he should go to Washington and apply for his patent.

Alec had to confess that he couldn't apply for a patent in the United States. His hands were tied.

"Why not?" the astonished Hubbard wanted to know.

When Alec had gone back to Canada in September, he had been just as nearly financially desperate as a man could be. He hadn't had the time nor the strength to continue his teaching—except for the lectures at the Boston University—and he had poured all the Hubbard-Sanders money into the materials for his experiments and inventions, and he had next-to-nothing of his own. He had always been reluctant to charge his deaf pupils for lessons and—well, that was that! Under the circumstances that had arisen at the end of the summer, he would almost have died rather than ask Mr. Hubbard for anything more, so when he had reached Brantford, he had gone to a friend of his father's, the Honorable George Brown, a member of the Canadian Parliament, told him about the telephone, and asked if he would become a backer. George Brown and his brother Gordon had been interested, and had promised to make a proposition after seeing his specifications.

"And did they?" inquired Mr. Hubbard drily.

"Oh, yes!" Alec assured him. "In December I had a letter signed by both of them, and they agreed to pay me twenty-five dollars apiece for a period of six months, in exchange for half-interest in all my inventions patented outside the United States, and the payments are to begin as soon as the British patents are obtained. But I must not file in Washington before the London office has been cleared, else the British patents will not be valid."

Mr. Hubbard sat forward quickly. "After the patents are obtained? My dear Mr. Bell, who is to obtain these patents—and when?"

Alec colored. "Mr. Brown offered to attend to that, sir. He is going to London shortly—indeed, he had expected to sail before this, and said he would see to it personally!"

Whatever Mr. Hubbard's alert mind might be thinking of both the Messrs. Brown and A. Graham Bell's business acumen, he strove to conceal it. He was learning that he must deal tactfully with his son-in-law to be. But for all that he looked concerned, and when he discovered that Alec was turning over one-half interest in the patents of all his inventions for the munificent sum of three hundred dollars, to be paid in monthly installments for six months, he was appalled. "My dear, Mr. Bell, the thing for you to do is to prepare your specifications for filing in Washington immediately. Can't you see your appeal to your father's friends has done you no good? You aren't a penny richer than you were!"

Alec nodded wearily. "I confess I have been put in a very tight spot. But" —there was a flare of obstinacy in his face— "I have given my word. I will not file for an American patent until I hear from London."

Mr. Hubbard sighed. He knew Alec Bell well enough now to know that he could be "just as inflexible when he was wrong as when he was right," and that he would sooner forfeit all his rights and opportunities than go back on his word once he had given it.

At last Alec had a message from the Honorable George

Brown. He was sailing from New York on January 25. Perhaps he was surprised at the arrival of three gentlemen just before he went aboard his ship. Alec Bell introduced Mr. Gardiner Hubbard of Cambridge and another gentleman, a patent attorney. Mr. Brown was delighted to see them. They talked over the Brown-Bell agreement, and Mr. Brown solemnly promised to cable the very minute the patent was filed in London.

Even before that, Mr. Hubbard had persuaded Alec to prepare another set of specifications to be taken to Washington as soon as word came from London. Probably to be sure Alec neither delayed nor forgot to write them, he was invited to prepare them at the Hubbard house. The older Hubbards left him in peace, but Mabel was trying to cure him of recklessly squandering the hours when he should have been asleep (she never succeeded!), and had made him promise to end his work at midnight. When the hour came, Mabel leaned over the balustrade and called to him. He had promised!

He couldn't make her understand from below, so he ran up the stairs and lifted his head so that the lamplight shone upon his face. "Just this once, Mabel! Let me off just this once! I must finish—now! There's a hole in the patent, and I must find it! I must find it now!"

The girl hesitated, and Alec said later, "that night a girl of eighteen held the fate of the telephone in her hands," for he was so weary and miserable in every way that had she denied him he would probably have thrown the specifications out and never begun again. But she didn't. She understood his need, as always. She smiled, nodded, kissed him and said, "Go on back and mend your 'hole.'"

He went back to his work-table joyously, and within twenty minutes had found "and mended the hole." It was a good thing, as events of the next twenty years were to prove, that the specifications he wrote that night were so clear and so perfectly worded.

Then began the waiting for the Brown cable—the cable which would never come, for the farther he sailed from

America, the more qualms George Brown had on the matter. Talk over a wire? Carry human speech by telegraph? How could he have fallen for such a crazy, weird American notion? (He forgot that A. Graham Bell happened to be a Scot.) The best thing to do, and the dis-Honorable Mr. Brown did it, was to put the specification papers in the bottom of his trunk and conveniently forget them.

Time drifted by, and Alec Bell still refused to consent to offering the papers to Washington. Luckily Mr. Hubbard became exasperated and said a quiet word to the patent attorney in Washington. "It is no use waiting longer for Mr. Brown; just you put in the patent." And much to Alec's anger, the thing was filed on the morning of February 14. On the afternoon of the same day, Elisha Gray, an electrician from Chicago, came forward with a "caveat" for something he too called "an electric speaking telephone."

This almost simultaneous filing was to be the root of many bitter disputes, although it is hard to see why; a caveat is, as Mr. Gray acknowledged, "the description of an idea never reduced to practice," and Alec's specifications were the blueprint for a completed instrument. Alec was summoned to Washington and granted the patent.

"A birthday present!" he exclaimed. It was his twenty-ninth birthday, just a year and a day since Professor Joseph Henry had shot out his emphatic order to get the necessary knowledge to perfect his telephone.

Perhaps it was fortunate that Alec did not have all the technical knowledge some other hopeful inventors possessed. One of them, a Moses Farmer, later cried out to Thomas Watson with tears in his eyes, "If Graham Bell had known anything about electricity, he would never have invented the telephone. He would have known the thing was impossible!"

Just a week after the granting of the patent, the glorious, unbelievable thing happened. Alec and Tom were making ready for a night's experiments with a new and improved transmitter which Tom had just finished. He described it as "a transmitter in which a wire, attached to a diaphragm, touched acidulated water contained in a metal cup, both included in a circuit through the battery and the receiving telephone. The depth of the wire in the acid and consequently the resistance of the circuit was varied as the voices made the diaphragm vibrate, which made the galvanic current undulate in speech form."

They filled the cup with diluted sulphuric acid and connected it to the battery and then to the wire extending between the laboratory and Alec's bedroom. They knew this new transmitter was good, but they expected to have to experiment with it half the night. Neither of them had the tingling moment of anticipation when Alec used to cry out, "Mr. Watson! We are on the verge of a great discovery!"

It was just routine. Tom went into Alec's bedroom, closed the door and lifted the telephone receiver from the bureau. Suddenly in his ear sounded an imperative call: "Mr. Watson, come here! I want you!"

Tom dropped the receiver and flung himself out of the room, yelling, "I heard you, Mr. Bell! I heard you! You asked me to come! What—what is it?"

Alec had spilled some of the sulphuric acid from the cup on his trousers, but the fact went completely out of his mind when Thomas Watson's words made their glorious impact. The telephone had spoken! It had spoken!

They stared unbelievingly, first at the wonderful transmitter and than at each other. Then, half laughing, half crying, they tested the telephone over and over. There was no mistake, no disappointment. Their words came beautifully clear.

Finally, when they could think of no more intelligent messages to call to each other, they began to recite, "One—two—three—four . . ."

Back in October, 1874, Alec had filed his first papers for American citizenship—but now, fumbling in his excitement for a properly dramatic flourish to end the first telephone conversation, "God Save the Queen!" said Alexander Graham Bell as his final words into the transmitter on that glorious night of March 10, 1876.

12

Reluctant Journey

"BUT YOU MUST EXHIBIT THE TELEPHONE AT THE EXPOSITION!"
Mabel was arguing. "This is your great opportunity! Papa
says all the great scientific men will be there, and Sir
William Thompson is coming from England to be Chairman of the Committee on Electrical Awards, and the whole
country will see and hear your telephone!"

The contrary expression on her fiancé's face was discouraging. "It isn't ready yet! It needs more tinkering—
more changes. It would be a mistake—I cannot leave my
classes at examination time."

Mabel Hubbard drew a long breath. She had been jubilant
when Alec brought her the news that he had been right—
he could prove that he could send human voices over a wire!
And now, when she had the inspiration to place the telephone on exhibition at the long-heralded Centennial Exposition that would be held in Philadelphia this summer to
honor the hundredth year of the Independence of the United
States, and which would be the swiftest and most convincing way to get people to see and hear this new marvel—now
Alec was turning contrary, just as he had over the affair of
the English patents!

She curled her fingers into his hand and looked at him
with pretended bewilderment. "Did I understand what you
said, Alec? That it 'needs more tinkering—it isn't ready?'
But I thought I understood you and Mr. Watson and Papa

129

and Mamma and Willie to tell me it was working so nicely!" She smiled a beguiling smile, "Oh, Alec, do enter it now, just the way it is! Why, you might even win the Grand Prize! And Papa will make all the arrangements for you——"

"It will be a pleasure"—Mr. Hubbard had come in from the library—"and since I am the Chairman of the Massachusetts Education Exhibit, it will be the easiest thing in the world."

Alec's thanks were sincere, but he refused to promise anything. He must think it over. And he dallied so long that the date for the closing of the exhibits in the Electrical Department came and went.

"Well!" said Alec, and it almost seemed as though he was relieved. That was that! That was final!

But it wasn't final with Mabel Hubbard. She informed him that the telephone could be considered just as much an educational device as electrical. And so could the various telegraphs. He had consented to enter a display of Visible Speech symbols. Very well; why not the telegraphs and the telephone—all together? And when he tried to argue or protest she squeezed her eyes tightly shut. "I'm not looking, Alec, I'm not looking! I don't see a word you say!"

Finally he surrendered. The telephone could go to Philadelphia.

Why was he so reluctant to let his fellow scientists and inventors discover that he had actually achieved "electric speech"—"a talking wire"? Did he flinch at the thought of possible failure? The telephone was still so new and strange that even the two men who brought it into the world couldn't always predict its behavior, or fix it if something went awry. Or was he simply loath to let this special brain child leave his hands? Mabel knew him very well by now, and she exclaimed, "He would have been tinkering with the telephone until the end of his life if I hadn't taken it away from him!" It was a strange twist of fate that a girl who could never know even a whisper of sound should launch the telephone on its spectacular career.

130

With the telephone finally established in Philadelphia, and Mr. Hubbard there in charge of the educational exhibits, Alec turned his full attention back to his classes. It was June, and examination and graduation time was looming ahead. It was just as well that what he called "telephony" was well out of the way!

And then he received a disconcerting telegram from Mr. Hubbard. The Centennial judges were scheduled to reach the section where Alec's exhibits were displayed on Sunday, June 25, and Gardiner Hubbard thought it imperative for Alec Bell to be there. In fact, he should come a few days before.

"Very unfortunate time," said Alec unhappily. "Of course I can't go. I can't leave my school and teachers right at examination time."

Gertrude Hubbard pleaded; she cajoled; she argued. Alec Bell could be as adamant as the Rock of Gibraltar. Thomas Watson, Mr. Sanders, Willie Hubbard—all of them tried to make him see that here was the most golden opportunity of any inventor's life. That he must go to demonstrate and explain his invention to the judges in the scientific terms they would need to know. Willing as he might be, Gardiner Hubbard would be helpless on that score.

Alec Bell could be a hard man to help, on occasion. The only reward they got for all their arguments was his growl, "I am not going!"

Here was another impasse. Without his presence the telephone exhibit would be worse than nothing, and this obstinate Scot was insisting that the examinations of a handful of deaf pupils and their teachers was of far more importance than the cherished invention over which he had labored for more than two years. His friends looked at one another, shrugged, and acknowledged defeat.

But they hadn't taken Mabel into their considerations. She had no willingness to see the invention she had worked so hard just to get to the Exposition sink into inglorious oblivion simply because its inventor wasn't there. She knew

131

it was no use wasting time or temper arguing with a stubborn man, but she had a better plan. With her mother, she called at Alec's lodgings, and then had the Hubbard carriage drive to Alec's school just as the afternoon session was over. It was such a warm day, she told him, and she thought he must be tired working on his examination papers. . . . Wouldn't he like a little afternoon drive with her?

After nearly a week of being on the defensive with everybody, Alec found her gentle, innocent and loving presence very soothing. He climbed in beside her and really relaxed as they drove down the lovely, tree-shaded Cambridge streets, with the scent of the June gardens all around them. Not one word of reproach did the understanding Mabel utter —in fact, Philadelphia wasn't so much as mentioned; and it wasn't until the carriage drew in close to the Cambridge railroad station that poor Alec woke to the fact that once again his Mabel was winding him around her little finger.

"What does this mean?" he exploded.

Mabel smiled at him guilelessly. "It means that you are taking this train which is coming in to go to Philadelphia, Alec."

"I certainly am not—" and then he stopped to chuckle in triumph. "I can't go! I have no ticket—no valise——"

Mabel opened her own bag and silently handed him a ticket, while the grinning coachman drew out the bag Mabel and her mother had packed at Exeter Place. Aleck stared at them both grimly.

"I—am—not—going!"

And then Mabel flashed her secret weapon. She could not hear the explosive emphasis, but she could read it from his lips, and she burst into tears. "If—if you don't love me enough to do this for me," she managed to choke between her sobs, "I—I won't marry you!"

The train was pulling in. A very dour Scot snatched the valise, pocketed the ticket and boarded the train, looking back to vow that he would get off at the very next station!

In Alec Bell's present mood nothing about Philadelphia

had any appeal for him. It was hot—he always abhorred hot weather—and in addition the city was noisy and crowded. Mr. Hubbard met him and took him to what was grandiloquently called the Grand Villa Hotel—"a curious place," Alec told Mabel, "it consists of six or seven private dwellings united together."

He discovered he was to room with Mr. Hubbard's brother James. Apparently the experience wasn't exactly restful. Alec refrained from saying that his roommate snored, but he did lament that

> In vain I turned from one side to the other—I counted imaginary sheep—watched imaginary waterwheels turning round—but all was useless . . . I thought I might as well occupy my time by writing a letter to you—which I did. I was afraid of disturbing your uncle, so I had to write in the dark.

Having groped for some paper he'd left on the table, he found a stub of a pencil and tried to forget his discomfiture by sitting on the edge of his bed and writing several pages, but when daylight came it showed that "the pencil had only marked in spots here and there," and even he couldn't understand the results, so he improved the dawn by writing another letter. . . . "I was half-sorry I had come and wished myself back in Boston."

Unfortunately Mr. Hubbard had half-laughingly said something about "smuggling in" Alec's exhibits. That troubled his uncompromising sense of integrity. "If I find that I shall have to 'smuggle' my instruments in, I shall come right back and relinquish the whole thing."

Fortunately Mr. Hubbard was able to show him the official catalogue of the Exposition exhibits with the listing "A. Graham Bell: Telegraphic & Telephonic Apparatus and Visible Speech."

But he still worried.

Oh! my poor classes! What shall I do about them? You don't know what a horrid mean thing it is for me to leave them at this time. The University building too—closes on the 21st, so that some other room will have to be found in order to close the session. I can't bear to think of it at all—or I shall go right back. Then there is the expense which I cannot afford. Travelling expenses, Hotel, rent of room for classes, and in addition I have to pay Miss Locke five dollars for every day I am gone. You really don't know what you do when you make me come here!

But at the end he added a contrite "Don't be entirely disgusted with me—nor feel that I am ungrateful for all the kindness of my friends."

The next day, however, he was feeling better, excited over having met Sir William Thompson—it was good to hear "a good broad Scotch accent"—and Monsieur Koenig, the inventor of the Manometric Capsule. They "had a long talk on scientific subjects in the French language. He spoke French and I English—and got on nicely!"

But now another difficulty loomed. Mr. Hubbard had to leave Philadelphia. Who was there to help Alec demonstrate the telephone? The judges were to make their rounds on Sunday, June 25, because the Exposition would be closed that day. On Saturday morning Alec appealed to Willie Hubbard by telegram. Could he come to Philadelphia and help demonstrate the telephone the next day?

That particular weekend was the hottest one of a scorchingly hot summer, and Saturday was a horrible day. The visitors who attended the Centennial moved listlessly about, and many of them were actually prostrated. Those who did pass the rude tables with the "telegraphic and telephonic" exhibits passed by with dragging feet and glazed eyes. If they noticed the tall, very thin young man standing beside them, they thought only that he looked as ill as they felt, which was the truth. Heat always made Alec ill and weak, and he had a nervous headache and was wondering whether Willie would arrive and what would come of it all.

His heart really gave a leap when he spotted Willie's familiar face among the descending passengers on the midnight train. Over a late supper Alec rapidly explained what had to be done about putting up the proper wires and testing the telephone. The judges would be in the electrical section by nine on Sunday morning.

They were up almost before daybreak the next morning, and at the main building very early. They decided that the best possible placing of the receiver apparatus would be in the gallery behind the great organ, but the stringing of the wires was grueling work. The entire eastern end of the building was made of glass sections, and there were no ventilators in the narrow gallery. The hot June sun streamed in unmercifully, and they worked in the scantiest possible attire.

Even the judges wilted at their job, and their feet dragged. Stopping by the table just ahead of Alec's one of them remarked, "Gentlemen, this is the last we will judge today. The rest can wait."

Alec was almost too hot and sick to care, but if they stopped now there would be no second chance for the telephone. By tomorrow be would be on his way back to Boston, and he would not return.

The early history of the telephone was a series of curious "accidents." By "accident" Thomas Watson plucked a "frozen" reed; by accident Alec Bell spilled acid and involuntarily cried out for help. Now, perhaps, came the most unlikely accident of all: when the short, portly man who had been accompanying the judges looked past them and saw Alec, his round face was wreathed in smiles as he started forward, his hands extended. "Why, Mr. Bell! What are you doing here? And how are the little deaf children in Boston?"

He embraced Alec enthusiastically, and the judges gaped. How did His Majesty Dom Pedro, Emperor of Brazil, happen to be on intimate terms with an obscure inventor?

Seeing their struggles to conceal their curiosity, Dom

135

Pedro hurried to explain, even while he was saluting Alec again.

"I see this wonderful man in your Boston. I am on visit to your country, you know, and I hear of his school where he makes the deaf to talk. I am much interested in the little deaf ones in Brazil, so I go to see Mr. Bell's school, and we had a fine, long talk! But now, Mr. Bell, why are you here?"

Catching his breath, Alec explained about the telephone. The Emperor's face lit with interest. He caught Alec's arm. "Come, come, gentlemen, we must see!"

One by one they examined the transmitter, and Sir William Thompson and Alec's old Good Samaritan, Joseph Henry, lingered over it.

"Could we listen to it first?" one of them asked Alec.

He nodded. "It is possible to talk one way only. My assistant, Mr. Hubbard, will go into the organ gallery with one of you."

"I shall go with the young señor!" Dom Pedro spoke before any of the others found his voice. "This I must do. Show me, young man."

Willie looked at the stout, immaculate Emperor, and had his misgivings, but there was nothing else to do, so he led the way to the little gallery behind the organ. Once the terrific heat struck the Emperor, he forgot his royal dignity and promptly followed Willie's example of shedding clothes until "both of them were reduced to undershirts and trousers." Willie took up the receiver, waiting until he caught the faint tones of Alec's voice, and then handed it to the Emperor.

The sound really was faint, and the observant Willie noticed Dom Pedro's face clouding with disappointment. And then the Emperor dropped the receiver and stared at the young American dumbfounded.

"My God! It speaks!"

He seized the embarrassed Willie and kissed him with fervor, and Willie was grateful for the sound of approaching footsteps in the hall as the judges came hurrying up.

"I heard! I heard!" the delighted Dom Pedro proclaimed. This thing—it does speak! Just as your great Mr. Bell says!"

"What did you hear?" Sir William demanded, "I mean—what words?"

"The thing was saying, 'To be, or not to be, that is the question——'"

The judges exchanged quick glances and nodded. They had stood beside Alec and listened as he said the prophetic words into the transmitter. But they were still scientists. The thing had to be proved beyond any shadow of a doubt. They all took turns listening. They all heard. They took turns talking. They tested the wires and the instruments. They even tried calling from the floor below, trying to find some loophole, Willie shrewdly suspected, to pronounce it some kind of hocus-pocus. They tested it for three hours while Alec and Willie held their breath—for it wouldn't have been impossible for the infant mechanism to go wrong—but it had never worked better. Finally Sir William announced, "Gentlemen, this is the most wonderful thing I have seen in America."

✎ 13

Adventurer in Sound

EXHAUSTED BY THE COMBINATION OF THE INTOLERABLE HEAT and the tension of the morning, Alec and Willie were glad to escape from the Centennial Building, with its glass-walled Main Hall heated to oven pitch, and return to what Willie described as "our rooms in the so-called hotel." They had both been too excited and worn to eat much lunch, and neither of them had the desire or energy to do more than smile at each other and rejoice.

But in midafternoon a bellboy brought Professor Bell a visitor's card, and when he handed it over, Willie read the name "Elisha Gray."

Both men were in pajamas, not a dignified costume in which to receive a professional rival, but, as Willie observed, "the heat excused much," and Alec Bell asked the boy to bring Mr. Gray upstairs. Bell was suddenly grave and alert, and Willie himself felt a quick foreboding of trouble. He had seen and even quickly examined Gray's exhibits, one of which he had labeled a "telephone." The outstanding difference between the two inventions might be that Gray's didn't work, but Professor Bell had said the man was a practical electrician. Was it possible that with a very few changes Gray's "telephone" would perform as well as Bell's —or perhaps better? What was the meaning of his visit?

But when he appeared Elisha Gray was all enthusiasm for the Bell telephone. He congratulated his rival heartily and apparently sincerely upon his glorious achievement.

138

"You have surpassed us all, Mr. Bell, with your wonderful invention, and I for one, would never have thought of producing electric speech by such means as you have used. I most unreservedly give you full credit for this marvel, and I have come to assure you that I have no thoughts of putting any claim before yours in this field!"

He was so generous, so frank and openhearted about it, that both his listeners responded to him immediately; and when he left, Alec and Willie looked at each other and relaxed. "We talked the matter over," Willie wrote, "and both felt great relief, for we felt that his action had removed all questions."

Once he was back in Boston on Monday, Mabel was justifiably thrilled by all Alec's reports. "If you do get the Grand Prize," she teased, "you must give it to me, for it was I who made you go after it! Won't I look gorgeous with the immense thing hanging round my neck?"

Left in charge of the telephone exhibit, Willie had several minutes of panic when he was required to move it from the Main Hall to the Judges' Hall, but that was followed by a surge of delight and pride when he could telegraph the inventor that he had moved it and everything had remained in perfect condition. The judges had wanted another demonstration, and had been even more impressed than the first time and Sir William and Lady Thompson had dashed back and forth between the transmitting and receiving apparatus. The telegram concluded jubilantly: 'The assembled scientists got so so excited and made so much noise that the police thought Judges' Hall was on fire!"

Then someone had the exciting inspiration that Alec try to place one instrument in Boston, the other in Philadelphia, and telephone to Willie Hubbard there. Professor Bell himself was confident that the experiment was perfectly possible, but "telephony" was still such an infant that the three cities of Boston, New York and Philadelphia among them didn't possess enough spools of the proper kind of insulated wire to attempt it.

Before the Thompsons sailed for England Sir William came to Boston for the express purpose of seeing the Bell laboratory and to make more tests with the telephone himself. The more he saw, the more delighted and outspoken in his praise he became. Alec presented him with two of his most successful telephones, but they were unskillfully packed, and were completely out of order by the time he arrived home and wanted to show them off.

Both Alec and Mabel begrudged being separated again for the summer. Mabel went to Cousin Mary's cottage again, and there is evidence that Cousin Mary and Alec Bell still were not bosom friends. Mabel burst out in a letter to Lina McCurdy,

> *I feel so cross, ill-tempered and out-of-sorts, I think I'll inflict my crossness on you. I have had a cold since last Sunday, and Cousin Mary won't let me go out-of-doors, also, Alec hasn't, unheard-of thing, been here since Thursday, and I don't believe he'll come today! Oh dear, I do hate the man's ceremonious Scotch politeness! If he were an American he would be content with Cousin Mary's general invitation, but being a Scotchman he will take care to come late tonight to avoid coming to supper!*

Possibly the "Scotchman" had his reasons for ignoring the "general invitation." And by the end of July he was beyond the temptation of seeking Mabel in Nantucket, having betaken himself to Canada.

A very strong and tender bond always existed between Alec and his parents, especially, perhaps, between Eliza Bell and her son; but after being with Mabel, he found his mother's deafness a more awkward, limiting and tragic handicap than Mabel's seemed. Apparently the "long, long black speaking-tube" wasn't overly successful, and even Mrs. Bell's own brother carried on most of his conversation with the finger alphabet or in writing. Although she always did understand her son, Alec reported to Mabel that when he had something personal or intimate to say, he said it to her in the finger alphabet. "When I tell her about you, for instance, it would be decidedly inconvenient were I to articulate so that everyone else could hear what I say!" He went on,

> You cannot appreciate—as I do—what a blessing the miracle of lip-reading is—it seems to me a greater wonder every day. When I am with you, dear, and speak to you freely by word of mouth, I often forget that you cannot hear—I never do so with Mamma.

No deaf girl could ever have had a higher tribute. Reading it, Mabel's heart leaped within her. There had been times, particularly in the summer before, when she was struggling with the first onslaught of Alec's love, when she had been tormented by the thought that he might be offering it out of pity, but now she knew better. She had actually made him forget that she was deaf!

She read on with pitying eyes:

> It seems so hard and cruel that she should be shut in all by herself—when it is possible to acquire the art of lip-reading. Before I saw you, Mabel, and before I went to Boston, I acquiesced in my mother's affliction—for to her it is an affliction . . . But now it pains me, for I know how much—faith and perseverence in lip-reading, could have done for her. How many sorrows and disheartenments might have been avoided and how many opportunities have been opened to her. It is my great grief when I come home to see her quiet resignation "under the Will of God" because of an obstinate dis-belief in the power of lip-reading on the part of my father and myself!

Poor Mrs. Bell! Even as she sat in the warmth of the August air, Mabel shivered at the thought of what her life would have been if her parents hadn't been the exact opposite of Alec and his father, and held fast to an obstinate belief in the power of lip reading. She wouldn't have been able to respond to every word Cousin Mary spoke, or to enter

141

easily into all the summer activities of the young people at
Siasconset. She wouldn't have been reading a love letter
from a fiancé who was about to become a famous inventor!

One thing she especially loved about Alec, and was
eternally grateful to him for, was the way he talked to her
about the world of sound just as fully and naturally as
though she knew what he meant. He wrote her about the
musical orgy he and his cousin Aileen Bell had embarked
upon when they went in to Nordheimer's Music Store in
Brantford one day and "played through every single four-
handed piece they had," and in consequence he'd come
out with a wretched head and with "I Waited for the Lord,"
Mendelssohn's four-part songs and Bishop's glees all mixed
up and ringing in his ears together.

Alec was involved in telephone demonstrations at home.
Naturally he had taken his precious apparatus with him,
and tried it first between the Bell home and one of the
outside buildings, and all the delighted neighbors took
turns talking and listening. When there was no one to share
the experiment Alec telephoned to himself.

And then he made plans for the first "long-distance" call.
He talked it over with the Manager of the Dominion
Telegraph Company, Mr. Griffin, and that gentleman was
so interested that he gave his permission to use the company's wires between Brantford and the town of Paris, eight
miles away. Alec set up a telephone with a "triple mouthpiece" at the Brantford telegraph office, and then drove to
Paris with the iron-box receiving apparatus, leaving his
uncle, David Bell, stationed at Brantford to recite Shakespeare and direct some volunteer singers. It never seemed
to occur to him that the telephone could be used for ordinary
conversation. When he had connected the receiver and
listened for results, "There was a storm of crackling, bub-
bling noises, but faint and far-off there was also the sound
of singing and speaking."

Alec frowned. The telephone should—it had every right
to—work better than that. Remembering that they were

using coils of very low resistance, he telegraphed Mr. Griffin and asked him to substitute an electromagnet with coils of high resistance. Suddenly the voices were so clear that he could recognize the speakers. When it was his uncle's turn, Alec heard a cough, against his ear, and then the words he himself had spoken to Dom Pedro, "To be, or not to be, that is the question . . ."

He frowned, bewildered. He had understood that his father couldn't be present at this particular test. And yet— was it a trick of the telephone that had caused the cadences of his uncle's voice to seem so much like his father's? He had to be sure, so he telegraphed to Brantford once more. "Was it my uncle's voice delivering that passage from Hamlet, or my father's?"

The telephone had been playing no tricks. It had been Alexander Melville Bell at the transmitter, and his son had recognized the fact from a distance of eight miles—which seemed to everyone another triumph for the telephone.

Another test, from Brantford to the town of Mount Pleasant—a matter of five miles—was a clear success, but the Canadian authorities and newspapers were lukewarm about the whole matter; and while the neighbors had plenty of fun listening or taking turns talking and singing into the queer contraption, they went away laughing at the "cracked notion of Crazy Bell." "Yes, sir," one of them commented years later, "when I was a kid we thought he was a regular nut."

By the time October came Boston seemed to be beckoning to Alec on all counts. It was a city where he could work and study and find plenty of materials for his experiments. He had even lectured on "telephony" there and found eager and respectful audiences. And across the Charles river there was Cambridge and the Hubbards and Mabel—particularly Mabel!

Mr. Hubbard had been more than understanding. Back in the early spring Alec had been offered a professorship in a small college, and Gardiner Hubbard had been alarmed that he might be tempted to accept. He had felt impelled to give Alec his not-too-complimentary opinions:

"You are not like other men, and you must therefore make allowances for your peculiarities. I do not mean to praise your peculiarities, for they are a great injury to you! If you could work as other men do, you would accomplish much more than with your present habits—but you must overcome these habits of your own will, and not by rules imposed by a college faculty."

He was probably referring to the Bell habit of working close to dawn, skimping on sleep, or making it up by sleeping well into the day, as well as what he considered the unfortunate flaw in Alec's personality, his failure to carry the multiple and harmonic telegraphs through to a triumphant finish. Like all people who, by some inversion of their systems, really do work easiest and best at night, Alec was goaded by others who tried to make him reorganize his habits and "behave like a sensible person."

When Mabel hoped to curb his night work, she poked gentle fun at him by telling him she was painting his portrait to hang in his laboratory. When it arrived, the "portrait" was the lifesized picture of an owl. To please her, Alec had tried to reform his ways, but they were both obliged to admit it was a failure.

"I cannot think in the daytime," he confessed to her miserably. "As you know night has always been my time for study and thought—and I cannot think connectedly until all the thousand and one disturbing sights and sounds of the busy day have gone to rest. For months I have been trying to break up this habit of thinking at night. I have been attempting to do all my work in the daytime, doing violence to my own instincts in the hope of working a reformation—but the result is that I have become unable to do any serious thinking *at all*! Now that I want to do it at all costs, I cannot! I feel that my only hope of accomplishing anything is to recommence night work."

144

Looking at his haggard face and desperate eyes, Mabel released him from his promise to be "sensible."

Probably Mr. Hubbard never understood why or how a man could do his best at three o'clock in the morning, but he did come to the conclusion that if the world was ever to have the full benefits of any of his future son-in-law's inventions, he must have a full-time assistant who would have the understanding, the skill and the loyalty to get them completed and readied for public use. So he and Thomas Watson had entered into an agreement in September. Watson was to quit his job at the Williams Shop and devote his entire time to working on the Bell inventions, for a salary of three dollars a day and a tenth interest in all the inventions.

"I believe there is money in your patents, and that if you take Mr. Williams' man and work with him, or let him work steadily on one thing until you have perfected it, you will soon make a success," Mr. Hubbard advised him. "While you are flying from one thing to another you may accidentally accomplish something, but you will never perfect anything. If you did not neglect the main thing, you would have succeeded long ago." He laid a kindly hand on Alec's shoulder. "I like you so much that I want to see you make your life a success—and it will be unless you neglect all your opportunities and talents. I am your friend!"

He was, and Alec's usually too-proud spirit acknowledged the fact. He gratefully accepted Thomas Watson's new status.

Mr. Hubbard still apparently had much more confidence in the multiple and "harmonic telegraphs" than he did in the telephone, and he made it evident that it was the telegraph work he wanted to see carried on. Alec tried dutifully, but both he and his assistant had lost interest and faith in it.

He cried out to Mabel one day in dark despair, "when will this thing be finished? I am sick and tired of the multiple—and the little profit which arises from it. Other men work their five or six hours a day, and have their thousands a

145

year, while I slave from morning to night and night to
morning and accomplish nothing but to wear myself out! I
expect the money will come in just in time for me to leave
it to you in my will! Oh! how I long for a nice little home
of my own—and a nice little wife in it, and some time to
rest! Don't scold me dear, . . . I am sad at heart, and keep
my feelings bottled up like wine in a wine cellar—only they
don't grow any better by keeping!"

Mabel did not scold him. She comforted him as best she
could, and in this instance she was wiser than her father.
She sent Alec back to his telephone, and finally he and
Thomas Watson concentrated their whole attention on it.
They proceeded to devise an instrument with both a transmitter and a receiver. "And then," in Watson's words,
"Mr. Bell decided his baby had grown big enough to go outside and prattle over a real telegraph line instead of gurgling
between two rooms!"

The Walworth Manufacturing Company, which had an
office in Boston and a factory in East Cambridge, gave them
permission to use the telegraph line between the two buildings some night after closing time. Alec went to the Boston
office, and Watson hurried to East Cambridge, two miles
away. No one was in the factory except a very skeptical
night watchman, as Watson adjusted the instruments carefully, connected them with the telegraph wire, and lifted
the receiver expectantly. Mr. Bell should be calling. They
had agreed on the exact time. But there was nothing but
silence—uncompromising silence. Watson did everything he
knew, and a few things more. Still the thing was obstinately
dumb. Watson's heart sank. He was wondering if it would
be possible for him to get his job at the Williams Shop back,
and he really was on the point of telegraphing Bell that
the telephone might be a success as a speaking-tube, but
that was the end of its ability, when an inspiration struck
him. Was there another relay in the building connected to
this same wire? The watchman didn't know and didn't care,
but Watson persuaded him to let him follow the wire, and

sure enough, in the foreman's office there was the mischief-making relay! He promptly cut it out, and tore back to the telephone.

It was dumb no longer. Bell's voice came shouting through it: 'Hoy! Hoy! Hoy!" (Never in his life would Alec Bell say "Hello" into a telephone.) "Are you there? Do you hear me?" Watson not only heard him, but he realized that the frantic Mr. Bell was growing hoarse!

Both of them were so jubilant by the time they reached their attic apartment in Exeter Place that they entirely forgot there were other lodgers in the place, and did a gala Mohawk War Dance. Going out the next morning Mr. Watson encountered their landlady, who was wearing a definitely acid expression (the word is his). "I don't know what you two fellows are doing up there," she told him bitingly, "but if you don't stop making so much noise and keeping the boarders awake, you'll just have to quit them rooms!"

Remembering the unhappy fact that both he and Mr. Bell were behind in their rent, Thomas Watson tried to be soothing.

And now someone—was it possibly Mabel?—was inspired with a plan to make the telephone begin to earn its own way. Whoever it was, it certainly was not Alec Bell. For all his foresight and genius in some fields, he was almost helpless where business matters were concerned, but Mabel Hubbard wasn't. She had inherited her father's business abilities in full measure, as well as her mother's gentle understanding. With such qualities added to her own vivid personality and charm, she was a perfect person to steer the Bell affairs.

At any rate, Messrs. Bell and Watson soon found themselves embarked upon a series of "telephonic lectures." Alec would take a number of telephones and go to Salem, Providence or New York—wherever he seemed most likely to be welcomed; advertise a lecture, place a number of receivers in strategic locations in an auditorium—much like loudspeakers of today—connect the telephone to the wires of some obliging telegraph company, and proceed to lecture, interspersing conversations with Mr. Watson in Boston. The Watson part in the demonstation would begin with "Good evening, how do you do?" Then Tom would sing the few songs he knew—a wonderful assortment including the hymns "Hold the Fort" and "Pull for the Shore," "Yankee Doodle," "Auld Lang Syne," "The Last Rose of Summer," and "Do Not Trust Him, Gentle Lady."

"My singing was always a hit," he remembered later, "the telephone obscured its defects, and gave it a mystic quality. I was always encored!"

He deserved to be, for he performed under difficulties that would have daunted anyone else. Remembering the landlady's "acid expression," he dared not run the risk, at his end of the line, of shouting conversation or roaring songs without some protection against being heard. So he devised the first telephone booth, constructing it out of barrel staves and the blankets from his and Alec's beds. It was suffocating but soundproof!

As for Alec, he was a born lecturer. He loved being dramatic, had a wonderful stage personality, never used notes, and always held his audiences captive by his beautiful voice. Also he had a magic gift of enthusiasm and the ability to translate scientific terms into really enthralling and understandable English.

The lectures did not always run smoothly. Sometimes it seemed as though every hobgoblin came out of hiding to haunt the infant telephone and its inventor. In writing to Mabel about a demonstration he had conducted in Boston, with Thomas Watson at North Conway, one hundred and forty-three miles away, Alec said everything that could go wrong had taken the opportunity to do so. The telegraph wires had been injured by a cold snap, and "it seemed as if a cyclone had been imported express by telegraph for the occasion. It was all the more mortifying as quite a number of skeptical telegraphic people were present."

Mabel herself attended one of the lectures in Salem, going

with Mrs. Sanders. They were seated close to the platform, but not directly in front, so although Mabel understood a good part of what Alec said, it was disappointing not to catch it all. There was trouble with the connection, and Alec was half an hour late in appearing. The audience grew impatient and stamped and shouted. When Alec finally appeared and asked the audience to keep perfect silence so they might hear Mr. Watson clearly, "the room became so still," said Mabel, "that I felt it, and hardly dared to breathe."

This time Mr. Watson's voice came through so clearly that it reached to every corner of the hall and everyone laughed and cheered. "Only poor me," Mabel wrote to her mother, "I knew there was trouble at first, and I thought they were laughing at Alec's failure!"

In Providence, Rhode Island, despite a very heavy snowstorm, two thousand people gathered to hear Bell and his telephone, and in spite of the storm's interference, they had a wonderful and very convincing evening.

Only one person was troubled about all this. Gertrude Hubbard had been brought up in an era when gently-bred people scrupulously avoided calling public attention to themselves, and she confessed to her husband, "I heartily and earnestly disapprove of Alec's giving telephonic concerts. I think it undignified, unscientific, mercenary—claptrap, humbug, Barnum, etc. But—but I want him to show the capacities of his instrument, and with dignity quietly claim for his invention all that it can do."

By and large the lectures were a success, and they did bring Alec some sorely-needed money. His profits from the very first lecture amounted to eighty-five dollars. He owed his landlady rent, and he was in acute need of new clothes, but he promptly went out and invested the whole sum on having a little silver model of the telephone made for Mabel Hubbard's bracelet.

She deserved it. Here was a girl whose fiancé's time, thoughts and conversation all pivoted around something she could never enjoy. Most deaf girls would have cried out against it in anguish, and begged him to abandon it in favor of something she could share. Mabel Hubbard set herself to learn to be intelligent about his invention, and really goaded him into further experiments, and always she was eager for news of its advance.

Another thing the "telephonic concert" did was to introduce Alec as "Alexander Graham Bell." That, too, was Mabel's doing. "Be sure," she commanded him, "that when you lecture, the gentlemen who introduce you say 'Alexander Graham Bell.' No A.'s for me."

Somehow, in the progress of the invention and the lectures, their courtship also managed to progress. Their love deepened and strengthened. Often they would go for quiet sleigh or carriage rides, and when they returned at what Mrs. Hubbard considered a rather unseemly late hour, Mabel would lift a dimpling, flushed face and plead, "Oh, Mamma dear, please don't scold us. We really couldn't help it. We were talking, so we had to stop under every lamppost so that I could see what Alec had to say!"

ᔍ 14

Bride in England

ONE DISTURBING DEVELOPMENT SHOCKED AND HURT ALEC TO the core. There were beginning to be assertions that Professor A. Graham Bell was not the inventor of the telephone. The *Chicago Tribune* apparently began it with an article which said, "many of the Eastern newspapers are favoring their readers with sketches of Professor A.M. [*sic*] Bell as the inventor of the telephone, while the real inventor, Mr. Elisha Gray of Chicago, sits quietly at home. Mr. Gray's claims are incontrovertible, and have long since been established by science and the columns of the *Tribune*, long before Professor Bell was ever heard of! They are officially approved by the Patent Office in Washington, and they have already brought in large returns in money as well as reputation to the inventor."

Unfortunately, while Alec was still dazed and quivering over the *Tribune's* assault, he received a note from Elisha Gray asking his permission to demonstrate the telephone in a lecture he was giving. He had a copy of Alec's patent and could make up a telephone, which he would explain was Mr. Bell's invention and not his own. Would Mr. Bell telegraph his answer—collect?

To Alec this was like salt on an already inflamed spot. He telegraphed his answer, but he paid for it. If Mr. Gray would refute the libel against him published in the *Chicago Tribune* he would have no objection!

Back came a very frank letter from a disturbed and angry

Gray. He hadn't seen the libel and he didn't know what Mr. Bell was talking about. He would investigate, but meanwhile he wouldn't show the Bell apparatus under any circumstances!

In the end both men wrote frank and forthright apologetic letters, but the insidious seed had been sown, and for more than twenty years Alec, who lived by the straitest code of honesty a man ever had, was to be troubled by whispers and taunts that he had not invented the telephone.

But the joy that came in July of 1877 swept away all thoughts of Elisha Gray. Mabel Gardiner Hubbard and Alexander Graham Bell were married in the great bay window of the Brattle Street house. The wedding invitations, with a deep black border, looked as though the Hubbards were going into deep mourning over the event, but actually indicated the recent death of Mabel's Grandmother McCurdy. Mrs. Hubbard had actually suggested that Mabel's trousseau should be black, but the thought of his nineteen-year-old bride in black made Alec shudder, and he put his foot down firmly in that regard.

Because he meant very literally what he said when he repeated, "With all my worldly goods I thee endow," Alec gave Mabel on their wedding day a legally executed document investing her with all his shares except one, and all his rights in the telephone patent. His second gift was a large, beautiful cross of pearls.

When Thomas Watson arrived for the wedding, he hid behind a tree on the Hubbard lawn to put on the white gloves he had been too embarrassed to wear on the Cambridge horsecar.

Afterward, with more trepidation than she would have liked to admit, Mabel went to Canada to see her husband's home and family. Before she and Alec were halfway up the path leading to the house on Tuleto Heights on the Grand River, a tiny person darted out and broke a cake over the bride's head.

"That's an oat cake, May," Alec laughingly explained, "and my mother did it because it's an old Scotch precaution against the bride's ever being hungry in her husband's house. He turned to speak with his lips close against his mother's forehead, and Mabel felt a surge of pity as he said, "Mamma, this is my wife. This is Mabel! I know you are going to love her."

One wonders how Mabel, with her imperfect speech, not knowing even one letter of the finger alphabet, ever talked with her mother-in-law.

Alec's mother was so tiny and quick in all her movements that she seemed to Mabel like a little bird. His father, by contrast, was big and burly, but with the kindest eyes imaginable, and the same easy-to-understand speech as his son's. Also there were the three Symonds sisters, relatives of Mrs. Bell's, who had recently come to live with the Melville Bells; as well as Alec's uncle, David Bell, and his daughter and son—Aileen, a rather formidable-looking young woman, and Charlie, who was to become, although Mabel couldn't have guessed it then, her brother-in-law. Alec introduced them as "the Bells from Dublin." Mabel also met Alec's sister-in-law Carrie, with her new husband, George Ballachy, and their baby son.

So many new people all at once—all with a claim on Alec! But not for nothing had Mabel been reared in the best traditions of Cambridge, New York and Washington. She met them all with easy friendliness and poise, and the light in her husband's face showed that he was proud of his wife.

He showed her to Brantford and Brantford to her in the next few days. "We can't pretend to rival you in scenery," he told her as they walked hand in hand over the Bell property, "although we do have some very lovely views here and there. The great defect of this country is its flatness. Mountains are at a premium. Add to this the interminable forest. Go where you will in Canada, your horizon is bounded by trees; you always are in the midst of a clearing!" Mabel agreed with him. In all her travels she had never seen anything quite like this. Here and there were giant trees, and stumps of others all around, showing how very recently the land had been salvaged from the wilderness.

Alec took her to the edge of the sand cliff in the orchard and showed her the grassy hollow where he had spent so much time when he had first come to Brantford, still numbed by the death of his brothers and half expecting to follow them. At that her fingers tightened their clasp on his and held him with such unexpected strength that it seemed as if she was holding him back from the long-past danger, and he smiled back at her in understanding as he kissed her.

"You see, it is shaped so much like a couch we call it the Sofa Seat. This is my dreaming place! Miles and miles of country lie extended below like a huge map. The Grand River comes from the left, and see, it flows right at our feet. When I lived here I used to spend hours in the Seat. It was my custom in the summertime to take a rug and a pillow and an interesting book to this very cosy little nook and dream away the afternoon in luxurious idleness." Mabel found it impossible to imagine her husband idle.

The second stage of their honeymoon was a voyage to Scotland and England. From aboard the S.S. *Anchoria*, Mabel wrote home that no sooner had they got fairly aboard than Alec had his coat off, and was "tinkering with the ships' bells which were out-of-order!" He fixed them, too! A few days later she was telling her mother-in-law, "There are very few cabin passengers, and almost all of them are Scotch. Alec says they all talk broad Scotch, and that he cannot help falling into it himself! I only know I can hardly understand him half the time!"

Mabel flourished aboard ship. The sea air brought fresh color into her cheeks, and vigor in her step. The peculiar condition of her ruined middle ear made it actually impossible for her to be seasick. Alec wasn't seasick, but he did develop a strange malady which caused pain and numbness all along his side and down his leg. Mabel confided this to her mother, but she pleaded, "Please don't let Cousin Mary know, or she would think she was right in thinking I have married a 'broken-down invalid,' which is by no means the case!"

Hale and hearty though he would appear in later life, Alec was frequently ill as a young man, and all his life suffered from prostrating headaches. What no one, perhaps not even Mabel, ever seemed to realize was that he lived at such an intense pitch, throwing every ounce of energy into what he was doing, that he exhausted all his natural strength and even his nervous reserves, and easily fell victim to illness when the climax of an invention was reached or some crisis was passed.

He had an elastic system, however, and rebounded swiftly now, as always. Within a few days Mabel was telling his mother that the symptoms were practically gone, "and Alec is brimful of new ideas on many and various scientific problems which quite pass my comprehension." She added, "Here we are safely ashore in Glasgow, in Alec's 'ain countri,' and the poor fellow is wild about it."

A little later they were spending a week in a rather primitive fisherman's cottage on the shore. Alec thrived on the diet of oatmeal and fish, and Mabel stood it as long as she could before she said half-laughingly, half-wistfully, "Do the people in Scotland have anything to eat except fish and oatmeal?"

Her contrite husband gave her one searching look and promptly set off on a five-mile tramp to the nearest village to buy something more tempting to an American palate. Perhaps the young bride wasn't sorry when they left Scotland and journeyed on to London.

Although she must have thrilled with excited pride at the welcome Sir William Thompson, who was now Lord Kelvin, gave Alec, bringing his scientist friends to met this "great inventor from America," it was well she was blessed with a sense of humor and understood her Alec so well. Few nineteen-year-old girls would have been content to watch a honeymoon turn into a series of scientific meetings and lectures. The visitors were polite to her, but then turned to plunge eagerly into scientific discussions with her husband.

One thing Mabel did have in common with the scientists: she was deeply interested in Alec Bell. Watching his happy face as he described his doings and his plans, she gloried in the fact that these Englishmen had brought him the understanding and sympathetic companionship he had never known before.

However, Alec wasn't the only one who had understanding and sympathetic companionship. Mabel's beloved Mary True happened to be in London, having left the Horace Mann School to become the governess of Sir Willoughby and Lady Jones' little deaf daughter, and Mary was able to run in often, delighted at the fact that the pupil and professor she had brought together four years before were now husband and wife.

By the last of October Mabel broke the news of their new plans to Eliza Bell, who had evidently been expecting this English honeymoon to be simply a holiday:

> I have not written before, because I dreaded telling you of our plans, and wanted Alec to do it, but he has gone off, and when he returns he will be occupied with his lectures . . . Alec has decided to remain here until next summer. He says that in America there are plenty of men who know all about the telephone and its management, but here there is no-one but himself. Then there are the foreign patents. And some gentlemen here have agreed to organize a company to control the patents here, and they say they must have Alec's presence for at least the first year. He thinks he is more sure of a good income here than at home.

So they had taken a house in Kensington which seemed enormous to the American bride, with its four stories and seventeen rooms. She thought the dining room, with red wallpaper, oak wainscoting and a simple but handsome fireplace with attractive tiles around the grate, was the prettiest she had ever seen. There was a white wainscoted parlor, and a bedroom with a great bed curtained in delicate pink, and the windows were hung with dark red and heavy lace curtains "to prevent too much sunshine." Alec had fitted up two rooms as laboratories. Most marvelous of all, the house possessed a bathroom with hot water pipes, "which are *such* a comfort."

It was here that Mabel Bell embarked upon her domestic career, with the help of Miss Mary Home—who had been a servant in the senior Bell household—as housekeeper, and a little maid, Emma. Here she would meet and entertain all the prominent scientific men who came to consult her husband about his telephone and arrange for him to demonstrate it and to found an English Telephone Company.

And it was here that the Bells gave their first guest dinner, for Mr. Preece of the English Postal Telegraph Service, a gentleman who could be very important to young Mr. Bell. It proved to be a dinner complete with every catastrophe that could befall a bride.

In the first place, Alec had said how fond he was of hare, so I determined to give him a surprise and ordered hare, and had my dinner all planned; just at the last minute when the tradespeople were at the door, Alec said something which I thought meant he didn't want hare, so in despair I ordered leg of mutton. At two Alec discovered my plans, and trying to make things right said he wanted hare, so down I trotted and ordered hare! Then I would have floating islands, no matter if I didn't know how to make custards, I was fired with a laudable ambition to give Mr. Preece a good and not a usual dinner. Miss Home is a splendid plain cook, but her idea of dessert doesn't go beyond puddings, and a custard was beyond her. I told her I thought my Mother made custards out of yolk of egg, sugar, butter, cream and wine, and I was sure she didn't put it on the fire at all. Miss Home remarked she had never heard of any custard not made on a fire. I did not dare try the butter, had no wine, and finally made a custard out of my receipt book—yolks of four eggs, sugar and vanilla stirred in a jug on hot water over a fire.

Miss Home undertook to stir the custard while I industriously beat the egg whites, and flurried as I was, I

*let her go on until it boiled, which of course meant ruin.
We took four more eggs, and I stirred while she beat. I
forgot to stir, and the first thing I knew the milk was all
over the floor, and that meant ruin no. 2! We tried some
more milk, and after herculean efforts the custard got safe
through and was set to cool. How happy I was.*

It was half-past five; dinner was set for seven, so—
flushed, but elated by the final victory over the floating
island—Mabel went up to dress. Her hair was half braided
when a horrible thought flashed into her mind. Mr. Preece
had been invited for six o'clock. Pulling on a sacque, she
rushed downstairs, to be met by her guest coming up!
Alec, who had a feverish cold, was asleep, and she hated
to disturb him. Also Miss Home signaled to her, and then
informed her that the hare wouldn't be ready until the
expected hour of seven. Then someone trying to light a fire
in one of the laboratories filled the house with stifling
smoke. Poor Mr. Preece told his inwardly writhing hostess
he would go out and make a call.

Meanwhile, to make sure everything went right with the
table, Mabel set it herself. She told Emma she must remain in the dining room throughout dinner so that the
courses would move smoothly. The girl nodded. Mr. Preece
returned. Mr. Bell woke. The dinner was about to begin
when there was a summons at the door. Who should be on
the threshold but the doctor (who had been expected hours
ago, and forgotten), come to see about Mr. Bell's cold!

Well, the dinner had to go on, so Mr. Preece took his
host's place. Emma did not remain in the room, and the
courses went wrong. There was no dressing in the hare. Mr.
Preece carved—and somehow the gravy went all over the
table, to the surprise of the guest and the mortification of
the hostess. But that wasn't the end:

*I passed my tumbler for water. The pitcher handle was
cracked, and of course it chose that particular moment to
break! When Mr. Preece lifted it, the pitcher fell and*

*smashed the tumbler! Dessert came, and of course Emma
put the floating islands before Mr. Preece, and also of
course he did not see that the edge of the bowl was
fringed, and half of the contents went through the holes
instead of onto my plate! . . . I must have given Mr.
Preece a pretty idea of my domestic management!*

Mabel probably was in agony over the thought of having shattered her husband's prospects in England, but Mr.
Preece must have had a saving sense of humor or a for-
giving spirit, for he was one of the sponsors of the special
general meeting of the Society of Telegraph Engineers
held in London "for the purpose of welcoming Professor
Graham Bell to London." The members forgot to be staid
Englishmen and greeted the young inventor of the greatest
of all inventions—who was skilled neither in engineering
nor electrical science—with wild enthusiasm. And the British
newspapers which had once scorned the telephone as "the
latest American humbug" now proclaimed it "A Great
Invention"!

Bell was immediately in great demand. He lectured on
both the telephone and the education of the deaf, giving
telephone demonstrations wherever he went, to delighted
audiences, and entertaining influential men in the white
parlor and the red-and-oak dining room. Mabel became a
poised, charming hostess. Her one complaint was that she
had not had Alec to herself in the evening for a fortnight.

Alec really seemed to be coming into his own. Using
telegraph and cable lines he proved he could telephone
from England to Jersey, from Dover to Calais, from Dublin
to Holyhead.

And then came a breathtaking invitation. Sir Thomas
Biddulph, secretary to Queen Victoria, wrote that Her
Majesty had become greatly interested in Professor Bell's
much-talked-of invention, and desired Professor Graham
Bell to come to Cowes, where she was staying at Osborne
House, to give her a personal explanation and exhibition
of his telephone.

15

"A Prophet Is Not Without Honor"

THE PROFESSOR MADE PAINSTAKING PREPARATIONS FOR THE telephone's presentation at Court. If the Queen approved it, both the telephone and the Bell reputation would be established in England. He would give Her Majesty a real "telephonic concert," but even better than any he had given in America, for the telephone was now superior to the ones he had used at home. He engaged four singers to sing four-part songs who would be stationed in the town of Cowes; Mr. Preece was to be ready in Southampton to offer Her Majesty a bugle salute; and the sound of an organ was to be brought from London! Young Mrs. Bell ordered a new evening gown.

And then came the first disappointment. Sir Thomas Biddulph made it clear that the "invitation" had not included Mrs. Bell, and she would not be permitted to accompany her husband.

So, still suffering from the heavy cold that had so upset the schedule of Mr. Preece's dinner, Bell went down to Cowes alone to see the Queen. He stationed Sir Thomas and Lady Biddulph, as well as an American woman journalist, Miss Kate Field, who was to sing into the transmitter, at a telephone in Osborne Cottage, some distance from the main house, and hurried over to take his place in the Council Room of Osborne House itself. He was not happy to discover, when he made his last-minute tests, that apparently the lines from Cowes, Southampton and London were all dead; but there was no time to do anything about it, for the Queen was being announced, and Sir John Cowell's coattails could be seen backing into the room ahead of Her Majesty. She was bringing with her her son, the Duke of Connaught, and her youngest daughter, Princess Beatrice. All the gentlemen in the room except the guest of honor bent double as they bowed. Had he become too American to follow their example? Sir John presented Professor Bell to the royal family, and Victoria smiled graciously as he bowed slightly. Then she turned herself to Sir John.

"Sir John, will you be kind enough to ask Professor Bell to explain the device he calls a telephone to us?"

When the gentleman-in-waiting gravely looked at him and relayed the request, Alec Bell complied, speaking simply and easily, and the Queen's face lit with an interest which he was told afterward was "most unusual." Several times she asked questions—always of Sir John, who duly repeated them, giving the lecturer the feeling he must be an Indian who needed an interpreter!

Finally Alec spoke to Sir Thomas, made sure the tones were distinct, and then handed the telephone to the Queen. She listened first with astonishment and then with evident delight, and finally handed the instrument to her daughter. The Princess looked at it a bit doubtfully, and then shouted, "Are you there, Lady Biddulph?" which caused her mother to slap her playfully on the shoulder and laughingly beg her not to scream. "The Duke went to the other extreme, first whispering, and then talking in a nervous squeaking voice."

When Kate Field started her "Cuckoo Song," the Queen happened to be looking away. Unconsciously the professor from America broke all Court rules as he leaned forward and touched the royal hand, and then offered her the telephone!

An involuntary gasp went around the circle, and the

culprit was told later that none of Her Majesty's closest ladies or gentlemen would have dared do such a thing. "That comes of having a deaf wife," Mabel commented shrewdly when she heard of it. But the august Victoria only smiled benignly at Professor Bell, listened to the song and said she was charmed with it.

After the audience was over and Alec was preparing to go, someone brought him a message: "Her Majesty desires to know if Professor Bell will do her the great favor to show the telephone to her upper servants? She would like them to see it."

In the midst of this second performance, the lines from Cowes and Southampton suddenly opened and it was the servants who listened to the four-part singing and the bugle salute. Professor Bell hurriedly sent word to the Duke of Connaught, who came "and was perfectly delighted, but the Princess had gone to bed, and the Queen was taking off (or putting on) something, and couldn't come."

"Alec says the Queen was 'humpy, stumpy, dumpy,'" Mabel reported to her mother, "her ungloved hand looked like a washerwoman's, so red and coarse and fat; and her face was fat and a bit reddish,"—and then apparently the Queen's visitor had a charitable afterthought—" but in spite of all she looked very nice, she was pleasant and dignified. As for Princess Beatrice she was beautiful and every inch a princess, and he fell in love with her at once. The Duke he thought a very fine young man, he liked him very much."

Two days after the royal exhibition, Sir Thomas Biddulph informed Alec "how surprised and gratified the Queen was by the telephone." She wanted to purchase the two instruments that were still in place at Osborne House. Unfortunately Alec couldn't take her up on that. The two telephones he had used in his exhibition before Her Majesty were the only ones he possessed, and he needed them for his other lectures and demonstrations.

Nevertheless it seemed as though both the telephone and

Alexander Graham Bell were well on their way. Alec demonstrated that he could talk to miners in a coal mine, and to divers under water. His calendar was crowded with lectures, experiments (he was still working with "telephonic improvements"), meetings with the English Electric Telephone Company and journeys to Paris to see about foreign patents. Meanwhile Mabel was not content merely to be on the sidelines. She was reading scientific books and papers in German, and translating them for her husband, which wasn't a simple task, for although she could read and speak German fluently, her schooldays in Germany and Austria had not made her familiar with scientific treatises. She was also acting as her husband's secretary, taking dictation from his lips and writing his letters until he found himself a professional secretary.

Another thing that pleased and thrilled Mabel was the publication of the first Bell "biography." Both the Professor and his telephone were so much the rage that Kate Field, the American newspaper reporter (they used the dignified word "journalist" then) who had "covered" the telephone's presentation to the Queen and taken part in the ceremony by singing the "Cuckoo Song" into the Queen's ear, was inspired to write a booklet about "Bell's Telephone," which was sold by the English Electric Speaking Telephone Company. It was authentic, too, for Miss Field had taken great pains to borrow Professor Bell's own scrapbook, in which he had pasted every newspaper and magazine article he had seen since the telephone was first mentioned.

But early in the spring the first sinister clouds in the Bell sky began to appear. Rival telephone companies were springing up and announcing that Professor Graham Bell had no claim upon the English patent for his telephone.

"No claim?" demanded the aghast Mabel. "But why not?"

Alec explained it to her. "You see, May, under English law, if an invention is published before the patent is taken out, the inventor loses his rights to it."

"But nobody ever did—oh—" she broke off with a little

cry at a sudden memory. "Yes, I remember . . . Sir William Thompson did! When he returned home after the Centennial, he made a speech about it in Glasgow, and you were so pleased!"

"He spoke to the British Association," Alec nodded. "And then his paper actually was published in the British Scientific magazine *Nature*." He smiled, but his eyes were troubled. "I shall telegraph Sir William for his advice."

Sir William hurried to the rescue. He stood up in court "and swore like a gentleman." Of course he had lectured on the telephone, and of course his paper had been published. He had even shown the two instruments which Professor Bell had been too kind and given him in Boston. But he added triumphantly that, as everyone knew, they didn't work! Now how could any right-minded English judge and jury think those two distorted, broken telephones were the same successful telephones which Professor Bell had proved to everybody did carry the human voice practically anywhere?

The judge and jury solemnly agreed with him and decreed that Mr. A. Graham Bell had a clear and unshakable right to his English patent.

But other troubles cropped up. The telephones put out by the rival companies were too hastily constructed or lacked some essential parts, "They get out of order perhaps twice each month," Mabel said, "and then Alec is blamed. It is most provoking."

If either of the Bells held the illusion that, once Alec had proved the telephone was a practical thing and not merely an amusing toy, it would soon rank with the telegraph as an important and popular invention, the English certainly shattered the idea in the next few months.

Someone had suggested to Alec that he give the Electric Speaking Telephone Company his ideas about introducing the telephone and possibilities for its future use. He worked over his proposals for days, and then sent a copy of them to each member of the company.

He suggested that telephones be placed in hotels, hospitals, railway stations, fire-stations and other public buildings. The telephones should not be sold, but always remain the property of the Company. They could be numbered and rented out, and even if a very nominal sum was charged the revenue would be enormous. The subscribers could be identified by the numbers on the telephones, and bills sent at stated intervals.

He wrote on and on, carefully thinking out each step.

The telephone should have its own system of wires, like the telegraph wires, perhaps, although he personally thought they would be better laid underground, in cables. There would have to be "central offices," where a man on duty would connect the proper wires when one subscriber wished to call another. Even one city could be connected by wires to others quite far away. These would be long distance calls, and tolls could be charged for them. He spoke of the advantage of being able to talk to miners in coal mines, or divers working under the sea, and he predicted a time when one could telephone across oceans.

The letter rang with the voice of a prophet, and met the fate that usually awaits true prophecies: it was blandly ignored. Much later, when he met one of the members of the Telephone Company, Alec ventured to ask what had been his opinion of the letter containing the proposals. The gentleman searched his mind for a minute, and then smiled a bit wryly. "Ah, yes, I recall it now! Well, I rather smiled, you know, and then—I tore it up!" Later the American Telephone Company was to carry out these ideas to the letter.

But regardless of all disillusionments, the Bells had reasons to consider this particular spring, with the English fields and hedges near them breaking into young fresh green, the most joyous they had ever known. In April Mrs. Hubbard and Sister came to join them in London, and Gertrude Hubbard reported to Eliza Bell that "Mabel never was half so charming and fascinating, and moved about her

165

home giving directions and superintending everything with as much ease and grace as though she had been mistress of a home for years." She had told Alec that she was very proud of Mabel.

"Not half so proud as I am!" he had said fervently.

As for Alec himself, Mrs. Hubbard told Mrs. Bell she would be more proud of him than ever; he seemed very well, but she thought he might be content with his two hundred and nine pounds. Never again would anyone describe him as "tall and very thin!"

And then came the climax of the Bell happiness. On May 8, as Alec wrote his mother, "our little baby popped into the world." His letter sounded proud, but somewhat awed. He continually referred to his little daughter as "it." "What a funny little thing it is, perfectly formed, with a crop of dark hair, bluish eyes, and a complexion so swarthy Mabel declares she has given birth to a red Indian!" He hadn't the vaguest idea about a name, and had agreed that since it was a girl, Mabel should decide.

This was certainly fortunate for the baby, because the two suggestions he did offer were "Jemima" and "Darwina" —the latter in honor of Charles Darwin. Happily, Mabel shook her head emphatically over both. Someone had told her that "Elsie" was the Scotch variant for Eliza, so the baby became Elsie May, with the May partially for her mother and partially for her birth month.

Despite himself Alec Bell had been haunted by a fear he wouldn't confess. A few days after the baby's birth he stole into Mabel's bedroom, and standing behind the canopied bed, he blew a lusty blast upon a borrowed trumpet. "The child is quite all right," he told a friend in great relief. "Mabel never moved, but the little one flung out its arms and legs and shrieked in terror." Elsie May Bell was not deaf.

Mabel soon devised her own ingenious method for "hearing" her baby. She placed her in a bureau drawer made soft with pillows, and kept the drawer beside her on the bed, where the vibrations of the baby's movements at night easily roused her.

But while Gertrude Hubbard was reveling in her first grandchild and the Bells were sending breathless bulletins to Brantford and Cambridge over the baby's growing charms and progress, Mr. Hubbard and Mr. Sanders, guardians of the telephone's destiny in America, were facing a sudden crisis that could mean the utter ruin of the Bell telephone.

Thomas Sanders, with the help of several relatives, had launched the New England Telephone Company, against all the storm warnings of his friends. "You have a good leather business," they reminded him; "stick to it. Can't you see there's no future in this telephone?" But the infant company acquired three thousand subscribers for itself within a few months.

And then, amazingly, the Western Union Telegraph Company sprang into the field with its own "American Speaking Telephone Company," advertising that it would "supply superior telephones by the *original* inventors, Thomas A. Edison, Elisha Gray, and Professor A. E. Dolbear"!

What had happened? And why? Mr. Sanders and Mr. Hubbard looked at each other incredulously and set themselves to find out.

Well over a year before, Western Union had asked its chief electrical expert, Frank Pope, to find out all he could about the telephone. After investigating for months, Mr. Pope reported that he thought the instrument had a great future, and he advised obtaining the Bell patents, because there really was no other way to transmit speech. However, someone at Western Union wasn't convinced. He thought it was ridiculous that an unknown teacher of the deaf should discover something Western Union's own skilled corps of experts couldn't produce better. And what's more, he didn't believe it. So he summoned Thomas Edison, who was an independent inventor, Professor Dolbear and Elisha Gray, and commissioned them to find another way—any way—to make a telephone transmit speech. They failed, but nevertheless in December of 1878 the American Speaking Telephone Company was formed by Elisha Gray and his partner and the Western Union.

Mr. Hubbard looked grim. He had met opposition many times before. He knew what had to be done. For the tiny Bell Company to sue the Western Union—the largest, most powerful business in the United States—would be like a pygmy striking a giant, but it had to be done if Professor Bell's rights were to be saved, and everyone else's interests protected. Mr. Hubbard was an excellent fighter if the right was at stake, and he made swift preparations. He cabled Alec Bell the whole story. Would he please send the scrapbook with all the clippings that told the story of his "telephonic concerts" and the improvements in the telephone as he made them—would he send that back at once? And where could they find his correspondence with Elisha Gray?

The whole thing was so shocking and distasteful to Bell that it made him literally ill, and when he tried to retrieve the scrapbook from Kate Field he was brought up against another catastrophe. Apparently it had been too much work to copy what she needed from the book, so she had simply taken her scissors and cut out the clippings she wanted to use. They had all gone to the printers, and after that—who knew? Miss Field was sorry, but that was that. Did Mr. Bell mind very much? Alec stalked away from her home with a prejudice against women journalists that was to last for many years.

As for his correspondence with Elisha Gray, he cabled, he thought he had probably thrown it out. That, too, had gone to the four winds. But Thomas Watson knew his apartment-mate. He went back to search the Exeter Place attic and found the letters in an unemptied scrap basket.

Now everybody was ready except Alexander Graham Bell

himself. Mr. Hubbard sent an urgent cable that the Bell family must return to America at once. The case would be lost unless the rightful inventor appeared and testified.

Over in London, Alec Bell was grim. He was disillusioned, disheartened and disgusted. If this was the way supposedly responsible business concerns like the great Western Union acted, he wanted no part of it. Elisha Gray, who had given him such generous congratulations—And he'd had letters of praise from Professor Dolbear. . . . They could *have* the telephone! As for him, he would take his wife and baby daughter to Brantford, and find a teaching position somewhere in Canada. He sent a terse cable to Gardiner Hubbard: "I am not coming to Boston. Will go to Quebec."

And then he turned to Mabel. "I have lost the telephone," he said.

ᦛ 16

Washington Circle

THIS TIME MABEL MADE NO ATTEMPT TO INDUCE ALEC TO alter his drastic intention. She must have had her own very definite ideas about his plan to abandon the telephone just as its value had become so evident that a powerful concern was using unscrupulous methods to claim it as its own invention; and she certainly couldn't have sympathized with his plan for flinging away all his scientific achievements and recognition, and returning to teaching in what probably would be a very obscure school. But Mabel had two exceptional gifts for a twenty-year-old girl—she had the wit to know when to keep silent, and she could also see another person's viewpoint.

She knew that added to all the disillusionments and disappointments Alec had suffered here in England these past few months, the perfidy of the two inventors who had been so quick and generous in their praise and the greed of a powerful company like the Western Union would be absolutely crushing to one of Alec's intense honesty. She could understand why her husband wanted to shake the dust of both Britain and the United States from his shoes and try to build a different life in a new country. She was also clever enough to suspect that he was too far away from the issues to see the affair in its true focus, and that if the die really was definitely cast for the Canadian school, he would repent in bitter loneliness. But, in any event,

170

Quebec would be much nearer Boston and the excellent patent lawyers her father had engaged than London was.

So they sailed for Quebec, and Mabel's heart leapt at the sight of a well-remembered face when they docked, for Thomas Watson was waiting to greet them.

After receiving Alec's cable, Mr. Hubbard, Mr. Sanders and Tom Watson had held a desperate consultation with the lawyers, who had warned them very bluntly, "Unless you act quickly, and get Mr. Bell back to Boston immediately, he will be too late to save the situation, and he will lose all his rights!"

The trio had exchanged looks of consternation, and then Mr. Hubbard said decisively, "I think the only sure way to get Alec Bell here is for one of us to fetch him, and I believe you, Mr. Watson, are the right man to do it!"

Thomas Watson had his work cut out for him. Alec was pleased to see his old assistant, but he made it only too clear he was not pleased at his errand. Mr. Bell's face could set in very Scotch lines. "I want to make it very clear to you that I have grown very dissatisfied with the entire telephone business, Mr. Watson. I am not having anything more to do with it. I intend to devote my life to teaching. Will you please understand, and believe that I am serious about this?"

Tom Watson thought he understood only too well, but looking at Mabel Bell he caught the quick gleam in her face that told him he had an ally, and he plunged ahead, trying to convince his reluctant friend that the telephone was on the threshold of something really glorious, and that it still actually needed him.

He filled the inventor in on the latest improvements as well as the business developments. For example, there had been a very clear need for some small, simple device to summon a subscriber to the telephone when he was wanted by another. One simply couldn't stand and shout into the mouthpiece in the hope that the other party would be in the vicinity of his instrument and hear the yell! So the

171

ingenious Watson mind had gone to work. First he had tried a "thumper," and then a buzzer, but neither of them had been quite successful, and then he'd had one of his flashes of inspiration, from the very name of the telephone's inventor. Every Bell telephone was now equipped with a polarized bell.

Between them, Tom Watson and Mabel Bell convinced Alec that he couldn't walk out on his backers and the Bell Telephone Company right in their hour of need—that if he did he would be quite as dishonorable as Elisha Gray and Mr. Dolbear. Still, he was determined not to be trapped into remaining in Boston or tying himself up with the company.

"I will take my wife and child to my parents' home in Brantford, and then, Mr. Watson, I will go with you to Boston, on one condition. I must have assurance that the Company will pay my expenses to Boston and back again to Brantford, and will pay for my services while I am staying in Boston!"

Watson took a long, deep breath of relief. "I can safely promise you that they will be responsible for all of that, Mr. Bell!" he exclaimed fervently.

But Tom Watson's promise wasn't sufficient for the obdurate Bell. He refused to budge until a telegram had been dispatched to Boston and the Company's "Yes" had been flashed back.

Very wisely, Watson accompanied the Bell family to Brantford—he says himself he "didn't want to run the risk of losing Mr. Bell"—and he must have fumed with impatience when Alec insisted upon remaining at his parents' home for several days, but perhaps this time there was a special reason for the Bell contrariness. After they were really safely aboard the Boston-bound train, Alec Bell became actually ill, and when the train arrived in Boston he was hurried to the Massachusetts General Hospital for an operation. He filed his preliminary statement from his hospital bed on November 20, 1878, and Mr. Hubbard and the other lawyers did the rest.

Apparently the Hubbard family was not in Cambridge at the time, for when Alec Bell was well enough to be moved, he was taken to the home of Cousin Mary Blatchford, who commented drily, "he may have been a patient sufferer, but he certainly wasn't a silent one!"

The suit of the Telephone Company vs. the Western Union dragged on for a year, and when all the evidence was in, George Gifford, counsel for the Western Union, had the unenviable task of informing his clients that it had been proved beyond any shadow of a question that Alexander Graham Bell stood as the one and only inventor of the telephone, and advised a settlement with the little Bell Company, pooling the patents and interests of both parties. The Bell Company accepted, retaining four-fifths of the interests on the various patents, the other one-fifth being awarded to the Western Union.

This victory surprised and dazed Alec who had become accustomed and adjusted to the gallant little company's flounderings around in a morass of debt, and to wondering from month to month how the meager payroll was to be met. When the news of the Bell Company's triumph over the powerful Western Union broke, and its stock soared almost overnight to a selling value of nine hundred and ninety-five dollars a share, it left him absolutely dumfounded.

The Bells did not make their home in Canada, but neither did they settle in Boston or Cambridge. The Hubbards were no longer there. More and more, Mr. Hubbard's work and interests seemed to be centering in Washington, and it also appeared that the family, none of whom was very strong, might be better away from the rigorous winter climate of New England, particularly the eldest daughter (Sister), who was struggling against tuberculosis.

Indeed, Washington seemed to beckon to all of them. The Graham Bells found a home at 1500 Rhode Island

Avenue, and a year or two later the senior Bells had forsaken Canada and come to be near their son, choosing a house in the famous old Georgetown section of the city.

Mabel frankly rejoiced in the move. She was delighted with the beauty of the Hubbards' new home and happy with her family near her, and with the opportunity to exhibit her baby's rapidly growing charms, but her own household was developing several very disconcerting minor problems: Miss Home, their English housekeeper, and Elsie's nurse, Annie were constantly complaining about Southern cooking.

"Annie says 'she hasn't had anything proper to eat since she came here,'" Mabel confessed, "and I am distressed! Monday she had beef-steak, Tuesday chicken, today roast leg of mutton! The colored servants won't eat 'sheep-meat,' and Annie doesn't like stewed mutton, and none of them will take corned beef! The colored servants want pork all the time—the others don't. Altogether it is getting very uncomfortable! Annie gets frightfully cross if any of us take Baby out, and yet she complains of not having her dinner regularly!"

Just as Mabel was in the throes of trying to help her English servants adjust to Negro help and American cooking, and the Bell-Western Union suit was still at a very dubious stage when Alec's presence could be vitally important, he burst in upon Mabel one day afire with excitement. He had been invited to join an exploring expedition to the Arctic! "Only a few months—perfectly safe—such a chance for making experiments and discoveries determining the influence of the North Pole on the magnet, hundreds of miles from telegraph poles—no fear of induction from them! . . . Oh, Mabel, think, wouldn't it be lovely to see the sun above the horizon for all the twenty-four hours?"

Sister, who happened to have dropped in to admire her small niece, looked at him in horror. "Yes!" she said indignantly, "yes, go and leave Mr. Edison to steal marches

174

on you! If he goes, I would get a divorce from him if I were you, Mabel!"

"Well, I wouldn't," Mabel told her diary, "and he may go if he wants!" Perhaps, all circumstances considered, the idea of joining an Arctic Expedition rather appealed to Mabel herself!

Alec didn't go, but his mind was so teeming with ideas crowding to be born that he had little time for disappointment. One of his most insistent dreams was the concept of a "wireless" telephone. If one could send speech along wires by means of an undulatory current, might it be possible to go a step further and do away with the wires? What about utilizing light rays? He didn't know—but for Alec Bell, to wonder about a thing was to find out.

This time he found an eager assistant in Sumner Tainter, who had, like Thomas Watson, acquired his scientific and electrical education at the Charles Williams Shop in Boston. With the asset of their working knowledge of the telephone, their new experiments progressed rapidly, and by the early weeks of 1880 their "wireless telephone" was complete. They had used a plane mirror of microscope glass made thin enough to vibrate when a person spoke into it, and a beam of light reflected from the mirror was thrown into vibration and picked up by a lens, reflected and carried to a selenium cell, which was connected to a regular telephone used by the listener in the usual way.

On a sparkling February day Alec Bell hurried to the laboratory on L Street. Mr. Tainter had already taken the transmitting apparatus to the top of the Franklin High School about two hundred and thirty yards away. This was their big test. They had proved their invention worked beautifully indoors, but how would it behave in the open?

Bell glanced anxiously at his watch. He had promised Mabel he would stay at the laboratory only an hour today. He signaled Tainter and then went inside and lifted the receiver. At once a clear voice sounded in his ear: "Mr. Bell! Mr. Bell! If you hear me, come to the window and wave your hat."

175

Alec Bell darted to the window and waved his hat frantically. The wireless telephone was a success! The "photophone" had been born. It was an even greater marvel than the original telephone. If one could speak into a mirror and have his words carried by light alone to someone in a distant place, would there be any end to the things inventors could accomplish with light rays? The photophone was the key to a whole dazzling new world, Alec believed. He had to hurry back home as soon as possible and tell Mabel the glorious news of the photophone's success.

But even as he was preparing to leave the laboratory, someone brought him a message from Mabel. The photophone wasn't the only wonderful thing to be born that fifteenth day of February—he and Mabel had a second little daughter!

By this time Alec was more acclimated to small babies. He acknowledged that he thought the newcomer pretty, but he still had a rather weird taste in names, for he actually proposed calling her "Photophone!" Fortunately for the baby, Mabel had other ideas. She would have liked to call her Roberta, but Alec shook his head, so the little one was named Marian Hubbard, for the baby whose death had caused Mabel such heartbreak years before. They were doubtful at first about Mrs. Hubbard's reaction, but when her permission was asked, she sent a telegram, "Marian—with my love."

Twenty-month-old Elsie was enchanted with her sister. Whenever she entered her mother's room her first words would be, "Hush, hush, baby sister!" In Mabel's own words, "with raised, fat forefinger, she goes hunting for the baby till she finds her in the cradle or in mother's arms, and has given her her soft good-morning kiss. She is supremely happy when the little one is held on her lap for a minute or two, and wants 'more baby'!"

Meanwhile the babies' father was plunging avidly into new work with his favorite invention, the photophone, and as he probed deeper into its mysteries and potentialities,

176

Four generations: Elsie May Bell Grosvenor holds her son, Melville Bell Grosvenor. Her father, Alexander Graham, stands between them and Alexander Melville Bell.

John A. D. McCurdy pilots his Silver Dart over Baddeck Bay in the first airplane flight in the British Commonwealth, February, 1909.

The A.E.A.: Glenn H. Curtiss, F. W. (Casey) Baldwin, Dr. Bell, Lt. Thomas E. Selfridge and John A. Douglas McCurdy.

Photograph by Gilbert Grosvenor

The "Wheel-Kite." Boy standing in center is Melville Bell Grosvenor,
later President-Editor of the National Geographic Society.

Alexander Graham Bell inspects a kite in his Beinn Bhreagh *laboratory.*
Photograph by John A. D. McCurdy

The Bell-Baldwin hydrofoil H.D.-4 ("Water Monster") roars across Baddeck Bay at 70.86 miles an hour, September 9, 1919.

Dr. Bell experimented with solar stills to provide shipwrecked sailors with fresh water distilled from the sea.

his enthusiasm and excitement climbed to new peaks. A letter to his father contained surely the loveliest description an invention ever had: "I have heard articulate speech reproduced by sunlight! I have heard a ray of the sun laugh and cough and sing! . . . I have been able to hear a shadow, and I have even perceived the passage of a cloud across the sun's disk!"

A French scientist, Ernest Mercadier, suggested the photophone be renamed radio-phone, and if his advice had been followed it would have been the first instrument to bear the word "radio," but Alec Bell clung to photophone. It was the first invention to use radiant energy, including invisible infra-red rays. Since it was limited to a "line of sight," and was useless during a storm or fog, it wasn't very practical in that day, but Bell was right in thinking that it would open up an entire new world of science.

1880 was quite a year. Besides the birth of little Marian and the photophone in February, January had seen the eldest Hubbard daughter, Gertrude—always called Sister to distinguish her from her mother—marry Maurice Grossmann; and April brought another Hubbard-Bell romance when Berta became engaged to Alec's cousin, Charles Bell.

In September, "circumstances," as Alec laconically put it, "called us to Paris." The "circumstances" were the summons by the French Government for Alexander Graham Bell to appear in Paris to receive the Volta Prize of fifty thousand francs for the invention of the telephone. It was the highest scientific award of its kind, and had been presented only once before, to Rumkorff, the inventor of the induction coil. Alec always counted it one of the greatest honors he ever received.

Back in Washington, with the francs converted into ten thousand dollars, Alec established the Volta Laboratory, where he and Sumner Tainter, and a cousin, Dr. Chichester Bell, who was on the faculty of the University of London as a chemistry expert and who had come to America on a leave of absence, could work out their inventions and experiments separately or in partnership, just as the mood struck them.

While her inventor-husband was conjuring his brain-children into being in the Laboratory, Mabel was winning herself a place in Washington Society. Her spontaneous interest in everyone, her gift of being a creative listener, her ability to make strangers surrender their hearts to her and her experience of living abroad, all combined to make young Mrs. Bell a delight both as a hostess and guest in Washington. There must have been many women who envied her good taste in clothes and her complexion "like alabaster with a pink glow coming through," as one Washington woman described it. She had sparkling blue eyes and her smile was one of her greatest charms.

Mabel liked Washington and its gaiety. "Such nice dinners Papa has once a week," she said, "with distinguished guests." She had such a healthy sense of humor that she could delightedly repeat the ridiculous episodes that had resulted from her own deafness. At one dinner she asked the senator who was her partner to repeat a remark she had failed to catch, whereupon he immediately stopped talking and his hands flew about in gestures Mabel couldn't begin to follow. She was relieved when she saw her father making his way toward them with Dr. Gallaudet, head of the Gallaudet College in Washington, where the sign language was used exclusively.

"Good evening, Senator," Mr. Hubbard said pleasantly. "I observe you seem to be having trouble conversing with Mrs. Bell, so I have brought Dr. Gallaudet to interpret for you."

"That won't be at all necessary." The Senator bristled indignantly. "I think I know the sign language quite well!"

"Ah, yes, I see you do," Mr. Hubbard smiled at him, "but my daughter doesn't!"

And then there were the calls—so much a part of Washington life. Grace accompanied Mabel on her calling tours.

178

"She is an accomplished caller," Mabel reported with admiration, "perfectly at her ease, with dignified politeness. When she sits up so straight with such cold dignity, I feel quite afraid of her!"

On one of their calls Mabel noticed a harp in the corner of the drawing room, and asked their hostess if she played it. Whereupon the lady turned to Grace and remarked that she was afraid Mrs. Bell might not be able to understand her very well, but would she please tell her sister that if Mrs. Bell liked, she would play the harp and sing for her, only Mrs. Bell must forgive her for not being at her best, as her voice was a little hoarse from a cold.

Mabel Bell made an herculean effort and kept a straight face while Grace gravely repeated the message, and then she smiled and begged her hostess to play and sing! So she sat and watched the plucking of the strings, caught a stray word or two of the song and applauded at the end. "I enjoyed it so much," she informed the musician, "the music was so soft and low!" Once she began talking she seemed inspired to go on and on until Grace gave her a warning nudge, stood up and said very firmly that they really must go!

In the summer of 1881 a thrill of horror ran over the nation at the brutal attempt to assassinate President Garfield. The doctor was unable to locate the bullet and unless it was removed it could breed a dangerous infection and end in death.

The desperate situation turned Alec Bell's mind toward an entirely new field, electricity in medicine. It seemed to him that if an induction balance could be made, a kind of electric surgical probe that would react to the presence of the bullet when passed over the President's body, the operation for removing the bullet would be relatively simple. He conferred with the President's doctors, and they agreed it might well work.

It wasn't an easy thing to contrive. Both partners threw all their skill and energies into the problem. Washington has a reputation for hot weather, but that summer was torrid, and both men were sacrificing rest and sleep, working into the humid nights as well as through the stifling days, experimenting on bullets fired deep into joints of meat, making the apparatus as simple and safe and light for the suffering President as possible. At last the device seemed ready for the test, and was carried to the White House. There were five doctors present. The President's anxious eyes followed every movement made while the Bell hands adjusted the many wires over his body.

"Mr. Bell," he asked faintly, "will you please explain this instrument to me before you go on?"

Alec Bell did, gently and as simply as he could, and the President moved his head in weary consent for the test to proceed. For some reason it was not a success, and the wounded man was too tired to have it repeated. They tried again four days later, and this time Alec sent word to the White House that if the President's bed was equipped with a steel spring mattress it must be removed or the induction balance would fail. Perhaps the attendants were afraid to move their patient—perhaps they thought that the upper hair mattress would remove the danger. At any rate the Bell request was disregarded, and the only result was a faint buzzing, while the probe failed to respond.

And now the newspapers changed their stories from accounts of "the wonderful bullet-seeker" to cries of scorn and abuse. Professor Bell was simply a publicity-seeker; this probe was a hoax! It was an ugly report and it made bitter reading to a man who had toiled through a hard, hot summer trying to create something that might save a beloved and valuable life. In all his life nothing would ever hurt as much as this accusation.

There was one more tragic happening to crown that bitter summer. Instead of a joyous announcement, Alec had to write a heartbroken note to his old friend Sarah Fuller: "We have lost a little boy named Edward."

$\backsim$ 17

Little Ship Without a Compass

NOTHING COULD SAVE PRESIDENT GARFIELD'S LIFE, BUT THE
Bell induction balance and its sequel, the electrical surgical
probe, were a very far cry from being the hoaxes the
hysterical newspapers called them, and grateful doctors
from all over the world proved it. In the days before the
X-Ray the surgical probe was a safe and painless tool to
locate bullets and other metallic objects in the body. The
faculty of the famed Heidelburg University was so impressed
that it conferred the honorary degree of Doctor of Medicine
upon the American professor whose skill and ingenious
insight had led him to surpass physicians and surgeons in
their own field. The French Government made him an
officer in the Legion of Honor, and invited him to speak
before the Academy of Science.

But what Alec Bell always counted as the most glorious
honor of his life came on November 10, 1882, when he took
the Oath of Allegiance to the United States and received
his citizenship papers.

Mabel said he grew so proud he was rather exasperating
about it. "Yes!" he would proclaim, "you are a citizen because
you can't help it—you were born one, but I chose to be one!"

The little girls were beginning to grow out of their baby-
hood now. Perhaps naturally enough, even when they
reached a very observant stage they weren't aware that they
had a famous father, or that there was anything unusual

181

about their mother. They were taught it was only courteous to face Mother directly when one spoke—no calling or answering from another room, or addressing her behind her back—and no one in the family ever mentioned her deafness.

The first time little Daisy realized her mother was different was one night when she was ill with some childhood disease and Mabel came in to see that all was well. Daisy pretended to be asleep at first, and then as her mother turned to leave, she spoke. Of course there was no answer, and the child sat up and called. Her mother simply went on her way, and closed the door. Feverish and frightened at this seeming desertion, the poor child shrieked for her mother until she was hysterical. Her father heard her, came in, listened to the story, and quietly explained.

As they grew older, both children were delighted to be "ears" for their mother—listening at the receiver end of the telephone, or repeating conversation too far away for her to see, by the mere motion of their lips; although Mabel Bell had grown so skillful at ignoring her handicap and making others unconscious of it, that one day when Elsie was in her teens and planning a party, she actually asked her mother where would be the best spot to place the musicians who were to play for the dancing.

Her mother laughed outright. "Don't you think that's a funny question to ask me when I can't hear a note?"

Elsie's first memory of the telephone and her first insight into her father's fascinating work came together one day near Christmas, when she was summoned into his study. "Come, Elsie, Santa Claus wants to talk to you," he told her. Bravely his small daughter scrambled onto a chair and pressed to her ear the receiver her father was holding out to her. She spoke unbelievingly into the mouthpiece.

"Are you really Santa Claus up in Greenland?" she demanded.

The voice assured her he certainly was, and that if she would tell him what gifts she most wanted, he would bring them to her on Christmas Eve. Satisfied and delighted, she turned to her father with her great brown eyes shining. "Isn't that machine wonderful?" she cried. "Do you know who made it?"

His eyes twinkled. "I did!" he said.

He was a remarkable parent in several ways. For example he didn't believe in the usual punishments, and wouldn't allow them to be used. Instead of administering a slap or a spank, Mabel was supposed to stick a pin into the fleshy part of the culprit's thumb—just enough to give a prick of pain. Her daughters can still remember holding out their thumbs and "being very much interested in the performance," until their mother rebelled. Her nerves couldn't take the punishment, although it didn't bother her naughty little daughters.

Another of Elsie's early memories came when she was five, and someone came to the nursery to tell her that the angels had brought her and Daisy a baby brother, Robert. But alas, the angels came back and took him away again. "Poor little one," their mother mourned, "he tried so hard to live!"

Perhaps it was the loss of his two baby sons that turned Alec's thoughts toward another medical invention. He called it a "vacuum jacket." The forerunner of the iron lung, it was designed for helping premature babies, reviving half-drowned persons, and remedying other breathing problems. But his chief interest at this period (at least at the Volta Laboratory) was a project he was working on jointly with Sumner Tainter and his cousin, Dr. Chichester Bell: a means of improving the phonograph.

In 1877 Thomas Edison had produced his phonograph. It was a marvel, but it did leave a good deal to be desired. The records were clumsy cylinder affairs for one thing, hard to handle or store and easily broken; and the recordings, cut with a blunt instrument on metal foil in an up-and-down pattern, were not very clear, gave a spotty result and were by no means permanent. The Volta trio set out to produce something more reliable, and all three men worked on it from 1881 to 1884. Undoubtedly Alec Bell's special aptitude for working with sound was the reason for their triumph.

They called their device a phonograph-graphophone, because it was a machine for both recording speech or music and playing it back. That mouthful of a name was too much for everyday use, so it became popularly known as the "graphophone." The contraption looked exactly like a sewing machine equipped with several weird attachments, and in fact that is what it was! The recording mechanism was far superior to the Edison model, using a sharp instead of a blunt instrument, and making a lateral pattern. However, the real triumph lay in the records themselves. They were made of a hard, waxy substance, and instead of being bulky cylinders they were flat discs. They were light, easy to handle and much less brittle than the Edison cylinders, and the recordings were permanent. The inventors predicted that their discs could be used in the future as a substitute for written letters, since they could so easily be sent through the mail. Instead of reading a letter, one would simply play it on his Columbia graphophone!

The sale of the graphophone and record patents brought Alec Bell two hundred thousand dollars as his share of the proceeds—which opened the door to one of his most cherished dreams. In addition to all his scientific work and shuttling back and forth to Boston to testify in the telephone suits—for there were more lawsuits concerning his telephone—he had opened a small office in one part of the Volta Laboratory to study the statistics and needs of the deaf, and his research assistant, John Hitz, had been pointing out for some time that it really should be a separate organization. After pondering the matter, Alec went to his wife with the idea.

"I want to take this graphophone and record money, and organize a Bureau—the Volta Bureau, we'll call it—and its business will be 'the increase and the diffusion of knowledge concerning the deaf.' I'll create a trust fund for it. And another thing: I want to start a new variety of school for deaf children—a dual school, where we can try the experiment of letting deaf and hearing children mingle together, just as you were taught to share all the activities of your sisters and their friends."

No matter how much energy, time and concentration Alec might be giving his scientific matters and inventions, he was still the professor with an intense concern for deaf children. How he managed to do it Mabel never knew, but he would always find time to carry on his crusade for education for the deaf, making surveys, writing articles and pamphlets, giving outspoken lectures before teachers' conventions and other organizations—always preaching the advantages of lip reading and speech in every way he could devise. With his flair for the dramatic, he could devise a good many!

Mabel remembered that he had startled and really shocked an audience of teachers by his declaration, "Nature has been kind to the deaf child; man, cruel!" Then he explained. "Nature has inflicted but one defect; man's neglect has made him dumb and forced him to invent a sign language which has separated him from the hearing world!"

And he had told Mabel, "My interest in the deaf is to be a life-long thing. I shall never leave this work, and you must settle down to the conviction that whatever successes I may meet with, your husband will always be known as a teacher for the deaf."

Alec's characteristic thoroughness made him investigate the claims and merits of the sign language system. He even learned it, but Mabel had never forgotten her fear and revulsion at the first sight of "signs" when the deaf-mutes in Boston had been witnesses at the hearing for the Clarke School. Perhaps it touched her with horror at the thought of what she had so narrowly escaped, but even when Alec's exceptionally graceful hands were demonstrating the Lord's Prayer in the sign language, she shuddered and confessed, "I know some people think it graceful—I find it hideous!"

Ardent advocate of oral education though he was, Alec

Bell could claim Dr. Edward Gallaudet, head of Gallaudet College for Deaf-Mutes, as a personal friend, and he had such an intense sense of fairness that when he was asked to give a lecture, he often invited Dr. Gallaudet to go with him and "speak for the opposition—and then I publish the results of the discussion on both sides!" Gallaudet College had been one of the very first colleges to honor Alexander Graham Bell with an honorary degree for his work with the deaf.

Mabel surmised that Alec's "dual school" idea was no sudden stray thought, but a plan that had been crystallizing in her husband's mind for some time, and that very shortly the school would open its doors. And so it did, first on Seventeenth Street, soon moving to a narrow house at 1234 Sixteenth Street. The sign advertised it as "Mr. Bell's Primary School," but he himself referred to it as a kindergarten. The deaf and the hearing pupils had their classes on separate floors, but they played together, and made friends with one another easily. Elsie and Daisy were both enrolled, and Elsie learned to read there, and their father took time out from his scientific work to teach. He had a very personal feeling for his pupils, just as he'd had in Boston, and when one small boy fell ill with the measles, he took the time from what must have been a crowded day to write to the child's mother, "We have missed little George very much. Indeed it quite spoiled our pleasure moving into the new school—not to witness his pleasure in all the pretty things we have for him." After cautioning the mother that it would be better to have George forget all the words he had learned than to strain his eyes, the letter went on, "When he returns we can do ever so much more for him than was possible in the 17th Street house. You have no idea how much the little fellow has won his way into my heart—and I hope he may be all well again very soon."

Bell's reputation as a teacher for the deaf was soaring high again. One day two very distraught parents came to his office. They had made the long trip from Alabama to Baltimore to consult a famous doctor about their small daughter. He had been able to do nothing, but advised them to go on to Washington and ask Alexander Graham Bell's help. This was a last-stand appeal.

It happened to have been an exceptionally busy day for Professor Bell, but he could never deny the need of a deaf child, so his secretary brought the family in, and Dr. Bell lifted the little girl to his knee. There seemed to be something even more strange and groping about her than was usual with most deaf children he had examined, and then the mother spoke in a low, strained voice, "Dr. Bell, this child—this child is blind as well as deaf! Is there any help?"

Professor Bell gave her one swift, startled look, and then he gathered the little girl into his arms and held her tightly against his heart; and as her amazed parents watched, her arms found his neck, and she snuggled close. She herself said years afterward that in that moment she instinctively knew she had found her first friend. Her mother said incredulously that it was the first time she had accepted or returned a caress from anyone outside her own family.

Over her head the professor was asking quick questions. What was her name? How old was she? How long had she been deaf and blind? Had the two come together?

"Her name is Helen. Helen Keller. I am Captain Arthur Keller, sir, of Tuscumbia, Alabama." Indeed, there was no mistaking the Southern drawl. "She is six and half now, and she had an illness at nineteen months."

Six and a half—Just exactly the age of his own bright-eyed, sharp-eared little Daisy!

"We still don't know what her illness was," Mrs. Keller began supplying the details in a faltering voice as she twisted her handkerchief. "The doctor didn't expect her to live, and then—and then he said she was perfectly all right! But one day when I was bathing her I saw that she didn't close her eyes when I passed the cloth over them, and she didn't turn her head when we spoke—or called—not even when we screamed!"

"How did you happen to come to me?"

"We heard of an oculist in Baltimore who had worked almost miracles, so we took her to him. He could do nothing," the Captain's voice saddened and impulsively his wife reached out her hand to clutch his, "but Dr. Chisholm advised us to come to you. He said he thought there was a chance she could be—educated—" the word seemed to hang in the air as though the father was afraid to voice it, "and that you would know how to do it."

Helen's busy little hands had been exploring her new-found friend's face, his hair, his eyes, his beard and his clothes.

"I'll take her now," Mrs. Keller started from her chair, "she must be annoying you. I'm so sorry."

Dr. Bell shook his head, "No, no, let her alone. I want to observe what she does."

Now the questing hands had found his pocket and brought out his watch. It was the kind that could "sing out the hours and minutes," and as Helen held it, he touched the mechanism and set it going so that she felt the vibrations in her hand, and wriggled and laughed out in glee like any happy child. Then apparently a new thought occurred to her, and she slid from his knee to explore the room, examining each object she reached.

Professor Bell's eyes followed her eagerly. At last he looked back at the tense, questioning faces of the parents and spoke with quick reassurance. "I believe your little girl has very great intelligence, and of course she can be educated! I recommend that you apply to the Perkins Institute for the Blind in South Boston. They have had no little experience and success in such cases, and I am certain will be able to provide a teacher for little Helen." He caught the child in his arms once more. "And please be very sure to let me know how she gets on, for I am greatly interested!"

This was no casual, comfortable dismissal. Somehow, for a minute, he and this child so completely sealed from the world around her had been in communication, and his heart which someone said "was larger than his body" was touched to its depth. But there was something more. The intent expression on her face, even the restless, roving motions of her hands showed she was trying to learn about the world that must be so baffling to her. He wanted to follow the progress of this child!

And he did. Two years later she was back again, accompanied by her joyous mother and the somewhat shy and awkward teacher from Perkins.

"Dr. Bell," Mrs. Keller introduced them, "This is Miss Sullivan, the teacher who has worked our miracle!"

As he held the radiant little girl upon his knee again, he knew that it was a miracle. The little groping, phantomlike child had been more than transformed—she was transfigured into an eager, alert, responsive, delightful little girl who had a surprising vocabulary in the finger alphabet.

He smiled at Annie Sullivan and gave her such understanding and sincere praise of her work that before the interview was over she had lost the combative spirit she usually presented to the world, and carried away the conviction that here was a man she could depend upon in time of need. She surrendered her heart to him for life. "He has the happy faculty," she commented, "of making other people feel pleased with themselves."

It was probably on this visit that Elsie and Daisy Bell were instructed to entertain Helen, and take her outside to play: whereupon they decided to show her their favorite spot. Daisy had purposely learned the finger alphabet, so she was able to "talk" to Helen and explain matters. Helen happily agreed to the proposal, so the three children joined hands and scampered off.

Later, strolling outside in search of the three children, Dr. Bell heard a cheerful call. "Here we are, Papa! Look up here on the roof!"

Raising his eyes, the horrified man beheld his two little daughters with Helen between them happily perched on the roof of Grandfather Hubbard's stable! His gaze went to the ladder. How had a blind child made the ascent, and then taken the perilous walk along the roof? He went swiftly to the rescue.

Many years later, Elsie, as Mrs. Gilbert Grosvenor, told the story. "He was a very lenient father, but that time—Oh, we got an awful scolding for doing that—taking Helen up a ladder onto the roof!" She paused, and then laughter filled her eyes at the memory. "But Helen certainly had a good time!"

Between Alexander Graham Bell and Helen Keller there would always exist a fine and very tender bond. He contributed greatly to her education, was instrumental in making it possible for her to attend fine schools and Radcliffe College, and always followed her doings with intense personal interest. Helen never ceased to love him. She said, "When he found me I was like a ship without a compass or sounding-line, lost in a dense fog." To him she dedicated *The Story of My Life*, and one day, many years after his death, when she met a Bell granddaughter in far away Japan, she told her wistfully, "I am still hungry for the touch of his dear hand!"

ᔍ 18

Transplanted Bit of Scotland

PROUD AS ALEC BELL WAS OF ACHIEVING THE RIGHT TO proclaim "I am a citizen of the United States!" and of all that his citizenship stood for (although he was always irked because residents of the District of Columbia lacked the right to vote), there was one thing about his adopted country and city in which he could take no delight—the climate of Washington most of the year. In summer it could really bring him perilously close to exhaustion. For several years the Bells had been seeking a place to escape the heat, but neither of them wanted to find it in a "resort." They knew exactly what they wanted: an unfashionable little spot near enough to the ocean for sea-bathing; a place of cool hills where they could "build a small cabin beside a running brook."

In the summer of 1885 Melville Bell had a yearning to revisit Newfoundland. He had spent a few years there as a young man before he married, and his thoughts kept turning back. It would be cool there, and that asset appealed to his son. The Bells needed a holiday. Organizing the Volta Bureau and the School had been hard work, and grief and care had pressed close to Mabel. Berta—the bright, the gay, the spirited Berta—had died suddenly early in the summer, leaving two tiny daughters, Helen and Gracie, and a heartsick Charlie Bell. Maurice Grossmann, Sister's husband, was dead after a tragic illness, and it was becoming evident that Sister was fighting a losing battle against tuberculosis. Even if it was just for the sake of their own little girls, the Bells must find a place where they could wait quietly for fresh strength and courage.

Newfoundland might be a very good place for that, and Mr. Bell senior would accompany them.

"Why don't you go by way of Cape Breton Island, Nova Scotia?" Mr. Hubbard suggested as they were discussing various routes, "and then you can take time out to investigate the coal mine at Caledonia, which happens to be a new investment of mine."

So instead of sailing directly to Newfoundland the Bells took a detour on a little steamer that carried them through the surprising inland sea—the Bras d'Or Lakes of Cape Breton Island. The captain told them that all in all there were four hundred square miles of water, forming a hundred bays and channels. The Bell party stood awed and breathless at the amazing beauty of the island. There would be tree-covered mountains apparently rising out of the water, and then would pass gentle slopes or sunny meadows— each new channel opening upon a vista which seemed more perfect than the last. And the air—untainted and fragrant with the scent of wild strawberries—made one feel clean and strong, ready to breath deeply, take a long, strong stride and leave fretting ways behind.

The steamer stopped at a tiny town named Baddeck, and Alec Bell was suddenly reminded that he had once come across a book by Charles Dudley Warner about this place, *Baddeck and That Sort of Thing*. He remembered Mr. Warner's descriptions. It was like a bit of Scotland; it was beautiful; it was cool! Also, it possessed "an un-hotel-like hotel, The Telegraph House, with flower gardens and welcoming lights." Glad to be off the cramped little steamer, Alec set out for a stroll around Baddeck, and with some idea of investigating the Telegraph House. On his way he stepped into a shop bearing the sign *Cape Breton Island Reporter*. The man who probably was the editor stood tapping his telephone with impatient fingers, and then turned a harried face toward his visitor, and in answer to his sympathetic inquiry, burst out:

"For the life of me, I can't imagine what's got into it! It worked wonderfully well this morning! And the worst of it is there's probably not a man this side of Halifax can attend to it!"

He broke off in open-mouthed consternation as Alec moved to the telephone, unscrewed the mouthpiece, dislodged a dead fly, and readjusted the mouthpiece within five minutes. Then he turned to smile at the bewildered editor. "I am Mr. Bell. Would you kindly direct me to the Telegraph House?"

The Telegraph House was refreshingly well aired and attractive. And it had pleasant aromas wafting from the kitchen. Alec had made up his mind. He hurried back to the steamer to bring Mabel to see his discovery. The place captivated her just as it had enthralled him. That afternoon the little steamer sailed on without the Bell family. What did the rest of Nova Scotia have to offer? They had found Baddeck!

They stayed on for three weeks, and each day had the growing certainty that they need search no more for their refuge from the heat and clamor and anxieties of Washington. They were reminded of Scotland, but Baddeck had a beauty and a deep satisfaction all its own. Somehow the Bells knew they had come home.

In three weeks they and the editor whose telephone had needed first aid established a companionship that would last many years. Arthur McCurdy was a prominent man in the little village, and Alec and Mabel surmised he was an exceptional one as well. Over their games of chess Alec confided his and his wife's hopes of finding a suitable summer home in Baddeck. Could Mr. McCurdy possibly help them? Mr. McCurdy could and did. Before the Bells left Telegraph House they had bought a small farmhouse about a mile and a half out of the town, from Grandfather McCurdy, and left directions for alterations.

And now they must think about Newfoundland. Melville Bell was sailing from Baltimore on the *Hanoverian*, and the Bell family were to join him at Halifax. It would be a pleasant voyage to Saint Johns, and a few peaceful days in Newfoundland would be a pleasant way to end the summer.

After the glorious sunshine of Baddeck, Mabel shivered in the dank chill of Halifax, and wondered if the sullen skies might be brewing something especially grim, but Alec assured her that Halifax was renowned for such weather, and that she might rest easy with a man like Captain Thompson in command.

But when a smotheringly thick fog closed down upon them, it was apparent that Alec was far from easy himself. There was a very real danger that they might approach too near the dangerous shoals of Cape Clear. He spent many hours on the bridge with Captain Thompson, trying to raise an echo from the cliffs about the cape by cupping his hands about his lips and shouting at the top of his deep voice. He did get an echo from a sailboat, which drifted away.

At eight o'clock on the third morning out the fog suddenly vanished and Alec saw a sight that made him catch his breath in quick dismay. The rocky shoreline was looming dangerously close to the steamer, and even as he looked he felt a shuddering, grating jar, and a second later the *Hanoverian* careened and lurched upon the rocks. He knew from the sounds that water must be pouring through holes in her hull.

Half-dressed, frightened passengers were crowding up the companionways, but somehow Alec maneuvered his way back to their cabin. The childrens' nurse was on the floor, praying aloud. He caught her up, seized Mabel with one hand and Daisy with the other, and managed somehow to make the nurse understand that if she could control herself enough to bring Elsie up to the deck she would be safe.

Back on the deck a half-hysterical woman clutched his

arm. "Mr. Bell! Mr. Bell!" she sobbed, "are we going down?"

He gave her a gentle and reassuring smile. "Oh, no indeed, madam. *We have gone.* Everything will be all right now."

Leaving his family in the care of his father, Alec made a quick inspection of the facilities for getting ashore. What he found did not please him.

"There are too few lifeboats, and the life preservers have all been appropriated by great strapping men," he reported to Mabel in disgust. "However I am certain that the vessel is quite as safe as the crowded cutters—more so—and I believe it is best for us to wait until the boats can land the other passengers and return for us. Fortunately the sea is very smooth, so there will be no risk rowing to the shore."

With her eyes fixed on his face, and conscious of the strength that always seemed to emanate from him in times of crisis, Mabel nodded and managed to smile, even while her clasp tightened about her small daughters. Then she looked out over the stretch of gray water and at the cutters bobbing farther and farther from the steamer. Although the shore looked deceptively close, it was a good three miles away, and the wait seemed interminable before her straining eyes saw the boats begin their return trip. Then Alec and his father were suddenly busy fastening ropes around everyone to be ready to be lowered over the vessel's side. Five-year-old Daisy was so little and light that she was tossed from the arms of one sailor to the next, and, when Grandpapa Bell appeared she heard someone shout, "Here! Wait for that big fat man!"

Even when they were safely ashore at the fishing village of Portugal Cove the difficulties of the *Hanoverian's* passengers were not solved. The villagers behaved more like pirates than anything else. Rude and sullen, they wouldn't offer food or shelter from the penetrating early September chill until they had promises of high payment. Finally the crew brought food and blankets from the steamer, and the villagers grudgingly agreed to let the "intruders" make themselves comfortable on the floor of the various cottages.

Captain Thompson asked Alec to stand guard over the mail and the ship's money and valuables. Bell, having no liking for the behavior of the rough-looking men who kept peering in at him, took to cleaning and carefully reloading his revolver, all within plain sight of the open door, and was left undisturbed.

The next day everyone set out for Trespassy, eight miles away, where the British man-of-war *Tenedos* was waiting for them. Even in a cart the road was unspeakably miserable and rocky, and there were just two horses in all Portugal Cove to accommodate the *Hanoverian's* two hundred passengers, but when the lame, sore and chilled party finally reached the decks of the *Tenedos* no shipwrecked people ever had a more royal or tender welcome!

There was no Newfoundland excursion for the Bells that year. Once they reached Saint Johns, they found the first steamer back to Baltimore and home.

Alec and Mabel Bell must have needed all the courage and strength and quietude they had found in Baddeck to carry them through the next months, for they returned to Washington to find that a new and particularly ugly lawsuit had broken over their heads. A company organized in Tennessee, the Pan Electric Telephone Company, claimed to have discovered that a German named Phillipp Reiss had really invented the telephone in 1862, and that Alexander Graham Bell had merely stolen the idea and hoodwinked the Centennial judges into believing it was his own. Besides, the Pan Electric Telephone Company claimed to have uncovered the fact that he was guilty of an even blacker crime: He had known all about Elisha Gray's caveat, and somehow had connived with his patent attorneys in Washington to bribe an official at the Patent Office so that they could steal Gray's caveat and graft portions of it onto the Bell patent!

If anyone had known or tried to find out what it was Phillipp Reiss had actually accomplished (he had sent musical notes over a telegraph wire several years before, but had not transmitted speech) or had bothered to investigate the Gray caveat for an instrument that refused to work when it was finished, it would have been evident how baseless the charges were, but apparently no one did anything of the sort. Nine and one half years after the Patent Office had issued the patent for the speaking telephone to Alexander Graham Bell, March 7, 1876, number 174465, the Attorney-General of the United States was persuaded to authorize the United States District-Attorney in Memphis, Tennessee, in the name of the United States, to file charges against Alexander Graham Bell for "perpetrating the infamy of the most gigantic fraud of the century" and to file suit to cancel the Bell patents.

Seven years before, Alec had thought he was ready to surrender all his telephone rights rather than go through the sickening business of proving other men frauds and liars. But this was something different. When any concern dared to question his honesty and then went on to concoct a story charging him with having connived to bribe an officer of the United States Government, and steal from a Government office, then he would fight with all the fury he could bring to bear upon the matter—and when he was roused, this stubborn, clear-headed, logical-minded Scotch-American could fight!

It was a grim period. Even the little school, so full of hope for everyone concerned, had to be abandoned, because, in the face of the "unjust and cruel suit," its professor had neither the time nor the heart to devote to it. When it closed Mabel told her diary, "Alec feels his life is ship-wrecked, and that I am the only thing he has left in the world."

Alec Bell never enjoyed writing letters, but when the occasion demanded it, he could write letters in which every word was clear, precise and pertinent. Now he wrote and published a nineteen-page open letter to the Attorney-General, refuting all the libelous claims of his opponents and establishing his own rights. No lawyer could have prepared a better brief.

Eventually the suit and the charges were dropped, and the Bell name vindicated, but the thing had dragged on for years, and for more years there would be fresh lawsuits casting their sinister shadows over the Bells' lives. It seemed as though there would always be conniving persons or upstart companies eager to snatch the glory—and the reward —of the telephone out of its inventor's hands, by "embarking," as someone said, "on that glittering quest of trying to break the Bell patent."

Even after Chief Justice Morrison R. Waite of the U. S. Supreme Court read the Court's decision on March 25, 1888, announcing that Alexander Graham Bell was certainly the true and only inventor of the telephone, and his method the only known way of transmitting speech by electricity, the suits kept cropping up for years.

Few of his closest friends ever guessed the heartbreak the Pan Electric outcry of "fraud" and "theft" caused Alec Bell, or the toll it took from him. Once when Mabel happened to be away, he wrote her, "But for you I would live the life of a hermit—alone with my thoughts!"

No one except Mabel would have suspected the genial, smiling-eyed Dr. Bell, who seemed to radiate enthusiasm and well-being, of harboring such ideas. But Mabel Bell knew it was more than possible. No matter how false the charges might be, Alec still cringed at having been accused of fraud and theft against the United States. She knew he had an urgent need of something to counterbalance the bitterness at the back of his mind. Was it Mabel who suggested the idea of the Wednesday Evening gatherings?

Whoever may have had the inspiration, it was in the midst of the Pan Electric trouble that Alec sent out invitations for the first of what were to become famous in Washington as "Mr. Bell's Wednesday Evenings," gatherings of men who were interested in scientific affairs—or any other outstanding matters in the world—who came to the Bell home to discuss a wide variety of subjects, or just to listen, and then to feast on the "simple but hearty supper." The guests discovered that Professor Bell was interested in practically everything under the sun, and that he was skillful at drawing them into the conversation. No one ever needed urging to attend a second time. For over thirty years each Bell "Wednesday Evening" was an event for the men of Washington.

The Bells had shared shipwreck and the insidious maligning campaigns of the lawsuits standing together, but there came a time of peril and disaster that Mabel had to face alone. Alec was away the early January morning in 1887 when fire started through the house at 1500 Rhode Island Avenue.

Fortunately for everyone, Beckie, the little Yorkshire terrier who considered herself Mabel's especial guardian, was sleeping in her mistress's bedroom, for the door of the room was locked and the panic-stricken housekeeper forgot that she could reach Mrs. Bell by going through the children's nursery. But wakened and frightened by the clamor and perhaps the smell of smoke, Beckie leaped frantically upon her mistress and licked her face.

Sensing danger, Mabel lit her candle and opened her door to face the terrified housekeeper and the other cowering servants. The combination of the flickering candlelight and the frightened stiffness of the woman's lips made it difficult for her to see what she was saying, but one word Mrs. Sears was repeating was unmistakable: "Fire! Fire!"

Curiously, Mabel seemed to feel no terror. She laid a steadying hand on the housekeeper's arm. "Have you telephoned to the fire department, Mrs. Sears?"

The woman nodded quickly. "Yes, ma'am! The policemen who woke us all up did that immediately, ma'am."

"Then have the nurse dress Elsie and Daisy quickly. And wrap them well. It's a bitter night. And the rest of you leave as quickly as possible."

She helped the terrified nurse dress the sleepy little girls and saw them carried to safety in a neighbor's house across the street before she turned back toward the stairs leading to Alec's study on the third floor. By that time the firemen had taken possession of the house. There was too much confusion for her to understand what they were trying to say, but from their expressions and their attempts to remove her from the house she knew what they must mean, and she shook her head.

"I must go after my husband's manuscripts and books," she told them. "If we lose everything else, they must be saved!"

The fire really was dangerously close to the study, and warning tendrils of smoke were beginning to curl through the room as she caught up Alec's precious manuscripts and pointed out the most valuable books to the anxious, impatient men beside her. She moved about the room swiftly, pretending not to understand their protests, until the room suddenly filled with swirling, choking smoke. As she reluctantly allowed herself to be led downstairs she was pleading, "Put all your efforts into saving the contents of this room! It doesn't matter about the rest of the house!"

The little girls were waiting for her at a neighbor's upper window. Mabel gathered them close as she had on the morning of the shipwreck, and watched the flames leap triumphantly through the turret room that had been Alec's study and soar into the night sky, blazing with a myriad sparks. The little girls were delighted with the spectacle and applauded it, but long shudders swept over Mabel Bell.

It was a difficult fire to subdue. In the zero cold of the January night the water froze almost as it struck the house. One witness said that "icicles as large as a man's leg hung from the roof." Mabel felt that her heart was as frozen.

It must have been a bleak homecoming for Alec Bell the next day, even though he had been assured by a telegram from Gardiner Hubbard that Mabel and the girls were safe. His reaction to this telegram was a family joke for years: Receiving it in the early morning hours, he had examined his timetable and found he had four hours to wait for the next train to Washington, so he returned to bed and went to sleep!

At home, he found Mabel quivering with indignation at the accounts in the Washington papers. "Mrs. Bell," said the reporters, "was carried out of the burning house into the bitter January night clad only in her nightgown."

"I walked out—and it wasn't my nightgown at all!" Mabel exploded. "I had put on my beautiful new white cashmere coat! Nightgown indeed!"

But the house was ruined. Walking through the havoc late the next afternoon, Mabel felt a wave of cold terror, and her hand tightened on Alec's. The firemen might be proud of the fact that the fire had not gone below the upper story, but no room had escaped the floods of water. Everything had been touched by the sodden destruction. It must have crossed Mabel's mind that she and Alec had known shipwreck and persecution by false witnesses, and now fire and flood! Thanks to her, many of the precious books and a good part of the manuscript on which Alec had been working were safe, but the rest of Alec's study was a shambles, lying in ice two or three inches deep on the floor.

A tall young mulatto was trying to bring some order out of the chaos. Mrs. Sears had told the Bells she had engaged several men to help clean away the debris, and that she had placed one boy in the study because he seemed unusually intelligent. Alec Bell, unfailingly interested in others and courteous to anyone who was doing him a service, stepped forward with his hand extended. "Ah ha, here you are. Is this Charles?"

"Yes, sir!"

Charles Thompson never forgot the heart-warming smile and the cordiality of the tall, black-bearded, heavy-set man that January afternoon. "In all my eighteen years," he said later, "I had never felt such a handclasp! I loved Mr. Bell from that moment."

"I am Mr. Bell, Charles," the newcomer was saying,

"and this is Mrs. Bell." Still clasping the boy's right hand, he placed his left hand upon the youth's shoulder and gave him a long, appraising look. Something in the intelligent face pleased him. "Can you read?" At the eager assent, he added, "I mean, can you read handwriting? See here—" Stooping, he retrieved a bit of paper from the floor and held it out. Proud of his ability, young Charles read it without hesitation.

"That's fine," Alec applauded. "Now young man, you have the most important work of anyone employed in this house! Don't throw away any scrap of paper, however small, that has figures, writing or drawing on it. Put them all in a basket or box, and when you are leaving, bring them to my room."

"Yes, sir!"

The thrilled youth set to work, and the Bells moved on with their grim inspection. None of the three guessed what was to come of that brief encounter. Alec and Mabel Bell had lost a home, but that day Charles Thompson had stepped into their lives. He would become their butler, Alec's valet, and Mabel's bodyguard in her husband's absences, and journey with them all over the world. For the next thirty-five years he would be the most dependable, understanding, loyal and loving major-domo a household ever had.

∽ 19

Lady of Baddeck

THE BELLS FOUND A HAVEN IN A HOUSE ON NINETEENTH Street for a year or two while their new house was being built. Mabel rejoiced when Alec chose the site on Connecticut Avenue next to Charles and Grace Bell. Two years after Berta's death Grace had been the third Hubbard to become a Bell bride. Mabel understood the close bond between her husband and his cousin, and since Berta's and Sister's deaths she and Grace had felt a special need for each other. Also, it would be nice to have all the Bell children grow up together.

It was in the new Connecticut Avenue home that Alec decided to work on an apparatus which would refute Mark Twain's remark that nobody ever did anything about the weather. The demonstation of the contraption gave one Washington reporter the most astonishing interview of his career.

It began when the reporter telephoned and asked if he might have an interview with Professor Bell, and what time would be convenient.

"I'll be very glad to have you," the inventor said cordially. "Come as late as you like."

The reporter thought perhaps six or seven o'clock?

"Oh, no, no," Mr. Bell protested emphatically, "nothing before eleven—or twelve—or even one or two at night; then we'll be quiet, and I'll be very glad to see you."

The rest of the story is one of Mrs. Gilbert Grosvenor's most delightful tales about her father.

A rather amused reporter presented himself at midnight and the door was opened by Mr. Bell himself, clad in a long dressing gown. He welcomed his visitor heartily, saying, "I think first of all we'll go to the pantry and raid the icebox for something good to eat." Having finished this hospitable rite, the host beamed and announced, "Now we'll go down to the swimming tank."

"*Swimming tank?*" gulped the reporter, "but—but I didn't—I didn't know anything about bringing a bathing-suit——"

"Oh, you don't need a bathing suit," the astonishing Mr. Bell assured him. "There's no water in the tank. It's just an experiment I am making in cooling the house. It seems to me it is just as sensible to cool a house in summer as it is to heat it in winter!"

He led the way down the back stairs of the annex into a concrete tank, at the bottom of which was a miniature study equipped with a rug, a writing table with a student lamp and two comfortable armchairs. As they settled down, the reporter exclaimed, "How cool it is!" Mr. Bell nodded as he lit a cigar. "Yes, you see I've had cool air brought down here. In the third-story bathroom I have installed an ice-box and an electric fan blows over the ice and sends the cool air into a canvas fire-hose outside the window down the three stories into the tank, so, as you notice, I have a cool room." He waved his hands. "It is actually a simple proposition!"

It might have been a simple arrangement, but the inventor neglected to say that it was definitely on the costly side.

The canvas hose wasn't the only external feature that distinguished the Alexander Graham Bell house from its neighbors. The inventor had the firm conviction that indoor bathroom piping exposed a family to sewer gas, so all the pipes for the plumbing had been placed on the outside.

Despite Alec's successful pioneer air conditioning for his own miniature study, with the first warning of Washington's summer heat and humidity the Bell family turned their faces eagerly toward Baddeck. They loved it more each year. The children reveled in the freedom and delights of the country; Alec Bell gloried in the proximity of the mountains, so like Scotland, where he loved to roam; and the whole family responded to the enchantment of the Bras d'Or with its wonderful moods and its little island and coves and harbors.

For perhaps the first time in her life Mabel did some of her own cooking, and reported to her mother-in-law, "I have done a little fancy cooking as a great and novel amusement, and have been preserving raspberries. . . . Wouldn't you like to look in on me and see the table set with our own butter and raspberry jelly?"

All the family had taken turns churning the butter. And when the little girls wearied of well-doing their father sat down at the piano and played "a fine marching tune" over and over to help them keep the rhythm and restore their flagging ambition. Neither Elsie nor Daisy had inherited his musical gift, and it was several years before they recognized their Butter Song as "Onward, Christian Soldiers."

But Mabel reported that the time came when "the work was almost all Alec's, and it convinced him that churning is a very barbarous and primitive way of making butter, and he is constructing a windmill for driving the churn! There is so much wind that he thinks it will work nicely."

Everyone was very well and very happy. Daisy looked like a little Micmac papoose. And then Mabel added her momentous news: Alec had taken unto himself seventy acres of land, "and proposes buying much more."

The property lies on a projecting neck of land that separates the Bay of Baddeck from the Great Bras d'Or, and commands magnificent views of the Entrance, the Little Bras d'Or, St. Patrick's Channel, with the mountains

*of Wycogomaugh in the distance. Beautiful dense woods
of fir, spruce, birch and maple cover the place from a
height about twenty feet from the water's edge at the
extreme point where they end in precipitous cliffs which
gives the name to the whole neck—Red Head. Alec is
very happy over his acquisition.*

He certainly was to buy "much more." But it would take
Alec seven years before he would be the proud possessor
of the entire tract he wanted. Convinced that Baddeck
required a different style of clothes, Mabel laid away her
mourning veils, much to Alec's undisguised relief, and
shortened her skirts to what her mother-in-law probably
would have considered a daring height. But at least she
could walk through the stony meadows and on the mountains without too much danger.

She had gone several steps further with Elsie and Daisy.
It was absurd, she thought, for them to be cumbered with
so many layers when they were romping and tumbling
about, getting into berry patches and playing about the
lake. She had taken them to the nearest store, and informed
the clerk that she wanted to buy suits for little boys, only
she wasn't sure of the correct size, but if they fitted her
two little girls that would be fine. Thereafter, while they
were at Baddeck, Elsie and Daisy would know the freedom
of the suits "for the two little boys."

She was also remembering the day of the party she had
decided to give so that the two girls might have the opportunity of meeting the children of Baddeck. As she
greeted the guests she had given the group a quick, observant glance, and made some excuse to whisk her daughters
away for a couple of minutes; then she stripped them of
their dainty stockings and party shoes—all the children
had come barefooted.

Mabel had been a charming bride-hostess in England,
and a delightful member of Washington's society, but it
was in Baddeck that Mabel Gardiner Bell came into her

own. She loved the clarity of the air and the sky, the mingled fragrances of spruce and wild flowers she had found nowhere else; she loved the forests, the mountains, the water, and the freedom and quietude which meant so much to her husband; but most of all she responded to the people. She had been happy in Cambridge and New York and Washington circles, but here with these sturdy, independent, intelligent country people the wealthy and cultured young woman was conscious of a bond of sympathy she had never know before. The identification became so strong that in speaking to or about them she always said "we," never "you" or any third person pronoun.

The feeling was reciprocal; between Mabel Bell and the people of Baddeck the unique tie was to grow and deepen with the years. They understood each other.

Pleased as they were with Baddeck and their newly-acquired property, the old McCurdy farmhouse didn't quite fill the Bell family needs. Once back in Washington, Alec drew up his own plans, made a scale cardboard model of exactly the kind of house he and Mabel wanted, and dispatched it to Baddeck to have the house built on the site they had chosen. It was to be called The Lodge.

Baddeck had brought them something more than a new homesite. They had a new member of the household, one who would be indispensable for many years. No longer the editor of the *Cape Breton Reporter*, Arthur McCurdy was Alec's secretary now. And the McCurdy family had acquired a new son named John Alexander Douglas. This child would one day play the leading role in Alexander Graham Bell's second most important project.

In the following years everybody in the Bell-Hubbard clans joyfully accepted invitations to gather at The Lodge —and not only relatives, but friends and acquaintances as well. Lina and Augusta McCurdy eventually built a house of their own in Baddeck Bay. Mary True was always welcome. Even Cousin Mary Blatchford was an enthusiastic

guest! She had evidently changed her opinion about Alec
Bell, writing feelingly about a time when she was overjoyed to see him. It was a night when

> *the rain it rained, and the wind it blew! The trees slapped*
> *and shivered, while the gentle lake boiled and danced, and*
> *all nature was in a frenzy. As for the house, there was just*
> *enough outside furniture—rockers, tables, flower-stands*
> *and awnings to swell the chorus.*
>
> *About midnight, Mr. Bell, who was tired with his work,*
> *desired a walk, so putting on his bathing suit, a close-*
> *fitting blue jersey, very becoming* [imagine Cousin Mary!]
> *and his slippers, he started up the mountain!*
>
> *Meanwhile, I got up to save myself the trouble of being*
> *blown out of bed. Putting my head out the door I saw*
> *Lina blindly reaching for her glasses and ear-trumpet* [Poor
> Lina McCurdy had grown painfully deaf]. . . . *We joined*
> *forces and swam downstairs, for the rain was streaming*
> *in the bay-window and the stairs were afloat! We were*
> *soon joined by Mabel, Miss True and Gardiner. . . . By*
> *3 o'clock it did seem as though Mr. Bell had had time to*
> *do his mountain thoroughly! Gardiner offered to go and*
> *search, but Mabel very sensibly remarked that if Gardiner*
> *could find his way up, Mr. Bell could find his way down!*
> . . . *Finally there was a tap at the window, and there stood*
> *Mr. Bell looking as handsome as Apollo with his grey curls*
> *wet and shining, and his white arms and legs. We gave*
> *him a joyous welcome, for he was so big and strong that*
> *it seemed as if the house would stand firmer with him in*
> *it! Then Mr. Bell, still in his costume de walk, raided the*
> *pantry and dealt out milk and cookies. . . .*
>
> *He is one of the most interesting men I ever knew, so*
> *approachable, and with a heart to match the size of his*
> *body!*

Bell enjoyed the wild beauty of his "Beautiful Mountain"
on all occasions. His daughter Daisy (Mrs. Fairchild),
recalls one winter when they stayed in Baddeck until February. "One morning after a heavy snowstorm, he told me

with great pleasure how he had walked up the mountain through the snowdrifts the night before, clad in his bathing suit and wearing snowshoes! He was really a very elemental man!"

He was, indeed. At night the people of Baddeck stood on the shore to gape at a tiny spot of light on the water which they knew was Mr. Bell smoking a cigar while he took his evening float.

Mabel had no share in these nocturnal and stormy ventures, but she did amaze her mother with such excursions of her own as a row on the lake at sunset one October evening:

> It was such a glorious sunset that I could not bear to go inside. Instead I rowed around to the harbor, and up it. Inside the sea lay like glass, every tree, every shrub, every color, was perfectly reflected, and it was hard to say just where the land ended and water began. The moon was shining bright and silvery on the water, yet the aftermath lingered on sky and land.

The family stayed in Baddeck far into the winter that year, and in December she was writing her mother,

> You will be surprised at my enthusiasm for the snow, remembering how I used to hate it, but snow here is a very different article from Washington or Cambridge snow. There it is a damp and chilly nuisance, here it is dry and nice to walk in. I used to be so cold going out, but here I have had several long walks when the thermometer was not far from zero, and have been as warm as toast all the time.

One thing about the young people of Baddeck caught Mabel's attention and troubled her. Once they had finished the little school, there was little opportunity for the young girls to do anything or learn anything that would make them independent, and Mabel had a stout belief that girls

209

should be self-reliant. She herself had been trained to develop her artistic talent.

She knew it wasn't feasible to seek out and encourage individual talents, so what would be the most practical thing for the average intelligent girl? Would it be sewing? What about a sewing school? The girls she approached caught her own enthusiasm. But if there were classes there must be a place to hold them. Mabel bought a little abandoned Congregational church, named it "Gertrude Hall" in honor of her mother, and engaged three teachers to teach her girls plain and fancy sewing. "Mabel is like her father, a born organizer," Alec told his mother, "there are about eighty girls under instruction, and already they have a market for their work in Montreal." But Mabel Bell had absorbed and enjoyed life in Nova Scotia for five years before she embarked upon her really important contribution to the life of Baddeck.

She got the germ of the idea in Washington. Sitting in the morning room of Mrs. Hale, wife of a Senator from Maine, she and several other women were making plans to form a "Washington Ladies' Club." She had been brought to the meeting by Alec's cousin Aileen, but she knew the group had been having weekly get-togethers for some time, to talk of books, art, current events and travel. Now they felt the time had come for them to organize a formal literary club.

And then Mabel had her flash of inspiration. "It seemed to me," she told Alec excitedly, "that if these women, at the heart of things, as they are here in Washington, with all the sources of information and entertainment offered them by a large city, felt the need of coming together and talking over things, how much greater must that need be, and how much greater the benefit to women like us in Baddeck, who are so far from things, and dependent on our own resources for all our information and entertainment!"

Her husband swiftly and heartily seconded her plan for a Baddeck women's club. He firmly believed that women ranked with men in their possession of brains, and had just as much right—and duty—to use them! They solicited Mr. McCurdy's help, asking him, out of his knowledge of Baddeck, to draw up a list of ladies who might like to pioneer in the idea. And the list he produced delighted both Bells. It included wives and daughters of the leading citizens as well as "just plain folks" and farmers' wives.

The summer of 1891 proved to be an especially hectic one, as they had a continual houseful of guests, and toward the end of the summer both Elsie and Daisy developed a particularly severe type of whooping-cough. It wasn't until all the "tumult and the shouting died" in early October that Mabel could summon her forces and with Mr. McCurdy's help send out her invitations to the ladies of Baddeck to gather at The Lodge on October 10 for what was to be a memorable day in the annals of the village.

When she actually saw guests coming on foot, and wagons and carriages discharging passengers at the foot of the path, Mabel was swept by a cold wave of panic, and begged Alec and Mr. McCurdy to support her at least by their presence. Both gentlemen emphatically declined. This, Alec told her, was her enterprise. They had helped her with the invitations and with suggestions, but this afternoon was hers alone!

Suppose her voice failed her? And there were people, she knew, who had difficulty in understanding her at first. Suppose, in her excitement, she failed to lip-read quickly? Her heart was still hammering as she took her place and faced forty-one pairs of inquiring eyes. But she bravely banged the little gavel Alec had given her, and made her brief speech of welcome. Then she turned toward Miss Georgianna, Mr. McCurdy's sister, and asked her to please read the paper which would explain the purpose of the club and suggest how it might be developed, asking for everyone to voice her opinion.

She was still nervous, but watching the eagerness that leapt into the listeners' eyes, and their swift responses, her heart soared joyously. She couldn't follow all the answers, but the dawning delight in their faces told her that the women of Baddeck were just as hungry to hear and do and learn as the women of Washington. They organized their club that very afternoon. Mabel would have liked to call it The Baddeck Club, but she didn't quite dare, for fear that the men of the village might think the women were taking possession of the town's name. So it was the Young Ladies' Club of Baddeck that came into being that day—the pioneer women's club of Canada.

✎ 20

Mabel of Beinn Bhreagh

ALMOST INSUPERABLE OBSTACLES CONFRONTED THE NEW women's club. Baddeck was a little village, sixty miles from the nearest railway station, and with no public library, but "The very difficulty made things more interesting," Mabel observed. "We had to take stock of what we had, or could do ourselves, and as the months and years went by, the amount of what we had and could do was surprising. And" —the flash of joyousness which was almost always present in her, darted out—"we had such a good time!

"There were no big maps, so we made them on wrapping paper and hung them on the curtains to illustrate current events. A doctor's daughter described a new surgical discovery with a skull loaned by her father, and drawings she had made. A young girl, too ill for other work, begged to copy our quotations in a book, and the collection was so good we thought of publishing it as a calendar."

None of the club members had ever thought of speaking in public, and Mabel, out of trembling experience, commented, "it took real pluck to begin."

Miss Georgianna McCurdy was elected president. Organizations for the people of Baddeck must be led by Baddeck folk, Mabel believed. She consented to be honorary president of the Young Ladies' Club, but never would be a presiding officer. Alexander Graham Bell, by popular request, had a hand in drawing up the Club's Constitution

213

and by-laws, which resulted in their being exceptionally clear, terse, and few.

Mabel was touched and profoundly grateful for one of the Club's unwritten laws—the rule that all papers should be read while the author remained seated. "It would not have been possible for me to take my share in the discussion which should follow papers, without the opportunity to look over the reader's shoulder, or get the paper sheet by sheet as soon as it was spoken."

The little club conquered its obstacles and flourished. It was ambitious. One of its programs listed for a month Icelandic Sagas, Folk Songs, a paper by Mrs. Bell on Home Industries, another on "My Personal Experiences in Bermuda" by Miss Caroline McCurdy, and Current Events. They would always include Current Events. Among the club's speakers were George Kennan, the author and explorer of Russia and Siberia, another famous resident of Baddeck; Sir Wilfred Grenfell of the Labrador Mission; E. J. Glace, the African traveler; Alexander Graham Bell and, later, Gilbert Grosvenor.

The Club—it is still spoken of with a capital C—made its own impact upon Baddeck. Someone read a paper on the need for sidewalks in the village, and a few months later the town acquired a plank walk between the old stone post office and Mr. McKay's store. Someone spoke up for an adequate and pure water supply, and the town authorities decided it might be time to investigate the matter. One Club member made forthright remarks about conditions at the village Academy, and Mabel suggested an idea just beginning to blossom in the United States, of a society made up of parents and teachers in the school. The members looked at one another with dawning enthusiasm, and Mabel wrote, "the result of that meeting was the formation of the Parents' Association, which had for its specific purpose the assistance, pecuniary and otherwise, of the Baddeck Academy." It was the first Parent-Teacher Association in Nova Scotia, and the pioneer in all Canada of what is now called Canadians Home and School.

The Young Ladies' Club was always alert to innovations. There were no registered nurses in Baddeck, nor indeed for many miles around, but when the Club heard that Lady Aberdeen, wife of the Governor-General of Canada, was sponsoring a new order of nurses in the Dominion, to be town-supported, the Club invited her and her husband to come to Baddeck and discuss the proposition with the Young Ladies' Club.

The Bells entertained their Excellencies at *Beinn Bhreagh* for luncheon before Lady Aberdeen spoke to the Club.

Charles had worked over the luncheon for days. He provided brandied fruits, creamed oysters, timbrals of chicken, breaded mutton chops, peas, potato croquettes, stuffed tomatoes, blue wing duck, jellies, sauces, celery and lettuce salad, cheese, olives, crackers, plum pudding, fruit, coffee and wine.

The children of Baddeck had made, Mabel noted, "a huge wreath ever so many yards long, that was hung in heavy graceful festoons on poles at the foot of the Court House Lawn. And a pretty little girl presented the great lady with a bouquet of sweet peas as big as herself!"

Despite the heavy luncheon, Lady Aberdeen evidently spoke with enthusiasm and conviction of her "Victorian Order" of Nurses, so named in honor of Queen Victoria's Diamond Jubilee. Before the afternoon was over the Club had voted to sponsor a Victorian Nurse for Baddeck, whose services later proved so successful that Mabel said, "Even Sidney, fifty miles away, desires to borrow our nurse!"

No wonder that after nine years Mabel could look proudly at the members of her Club and say, "See what we have done. In the first place we have lived. Lived, not vegetated—lived when even our best friends, the very founders themselves, doubted the possibility. Lived and thriven in the midst of indifference, incredulity and passive ill will. Today we are alive and prospering, strong in the affections of our neighbors—and the respect of our neighbors, a recognized power in the town!"

On the Club's twentieth anniversary in 1911 she declared, "I, for one, can truly say that there has scarcely been a meeting which I have attended these twenty years from which I have not come feeling refreshed, stimulated, and with new friendliness toward my associates. To me indeed the Club has fulfilled its object of stimulating the acquisition of general knowledge and promoting sociability till amongst my dearest memories are meetings with the people of Baddeck."

Even fifty years after Mabel read her tribute, her Club would still be strong and still be counted a power, not only in Baddeck, but throughout Nova Scotia. Only one thing was altered: In 1922 the Club became the Alexander Graham Bell Club of Baddeck.

All her life, books and reading would be of high importance to Mabel. She went so far as to read in bed with a lighted candle on the pillow beside her. And once, at least, she was known to fall asleep while the candle was still burning!

"No one ever reads anything here except the Bible and the newspaper," Alec pointed out when Mabel began mulling over plans for a library in Baddeck.

"That's because there isn't anything to read," Mabel said with incontrovertible logic. "I think the older people are set in their ways, but the younger people could be influenced."

George Kennan agreed with her. It was he who actually organized the library, but Mabel contributed books and magazines, engaged a librarian and established the library in Gertrude Hall.

By 1890 Alec Bell had achieved his heart's desire of acquiring the whole of Red Head. When he had first stood on its tree-covered bluff and looked up at the mountain above him whose lovely contours reminded him so much

of the mountains he had climbed in Scotland in his boyhood, he had cried out to Arthur McCurdy, "I must own
that mountain! That beautiful mountain!"

It had taken seven years before he and the owners of the
four farms on Red Head had come to satisfactory terms,
but in the end the entire point of land was his and he
promptly renamed the area, *Beinn Bhreagh* (pronounced
Ben Vree-ah), meaning, in Gaelic, "Beautiful Mountain."

Now the Bells set about choosing a spot for their perma-
nant home in a way of their own. Alec had a substantial
platform built across the estate's hay-wagon, with a chair
on top for Mabel, who, day after day, gathered her skirts
in one hand and bravely scrambled up the little ladder.
Then Alec mounted up behind her, and John MacDermid,
who was rapidly becoming the Bells' right-hand man, proceeded to drive slowly around the estate, stopping every
so often at a signal from either the Professor or his wife.

If some people of Baddeck stood in stunned amazement
to see the man they had come to appreciate, and the lady
who had won their hearts, slowly wending their way over
their newly-acquired property in this fashion day after
day, who could blame them?

The Bells knew perfectly well what they were seeking.
They wanted a site with a lovely outlook from every
possible angle. It had to be near the water, within easy
distance of the forest, with a view, if possible, of the mountains. And they found the perfect place.

The house they eventually built had its back to the woods,
with broad, grassy terraces leading to the cliff that dropped
into the Bras d'Or. On the far side of the lake, which completely circled the house, were the gently rising slopes of
Washabuckt Mountain.

The house itself is large and rambling, with a friendly
atmosphere. It was a far cry from the "little cabin beside
a running brook" of the Bells' early dreams, but no little
cabin could ever have satisfied Alec Bell's love of space
and his need of utter seclusion and quiet at times—nor

would it have done for the hospitable Mabel and the needs of two growing girls. Large as the house is, it was overflowing most of the time.

Beinn Bhreagh became a mecca for everyone, but especially for the younger generation of Baddeck. And no one who experienced Mabel Bell's radiant friendliness ever forgot it. Shy young Annie Laurie said, "She received me with such kindly interest I forgot I was a stranger."

Another girl was awed by the fact that Mrs. Bell had helped her prepare for a "wonderful fancy dress concert. She spent hours draping and fixing my gown to get the desired effect of a French lady of the old school. Nothing seemed to be too much trouble."

Nothing was too much trouble for Mabel. She threw herself unstintingly and joyously into the parties and dances and theatricals, but could turn her attention just as completely to any girl or woman who brought her an unhappy problem. "She was so interested in us individually," one of them commented years later, "so sympathetic when we had sorrow, so happy in our joys!"

When Mabel's gardens were a riot of glory she would send a hurried invitation to as many of her girls and Club members as she could reach to come and pick as many flowers as they could, sending her own boat to fetch them and take them home. And when the Bells entertained unusual people—Sir Wilfred Grenfell, Helen Keller and Annie Sullivan or Professor Samuel Langley—the village was invited to meet them.

One guest, remembering Mr. Bell's contribution to the occasions, remarked, "How delightful it was to sit in the tower corner of the library and have him read or talk to us, or perhaps he would go to the piano and play so wonderfully, and then his delight was to have us sing! Mr. Bell felt the music so that the keys became almost alive under his touch!"

Even when she went abroad Mabel remembered her

Club, and and wrote long and sparkling letters for its meetings.

She had a genius for loving and understanding people, and there have been few instances of one person and a village sharing the feeling which Mabel and the village of Baddeck had for each other.

The townspeople offered an extraordinary proof of their affection by passing an ordinance giving her the right to a Canadian vote. She was the only alien ever to receive this privilege, and it must have pleased Mabel deeply, for even as a girl she had held the belief that every intelligent woman should have the right to vote, and a voice in the government of her country.

✍ 21

The Kites Fly High

WITH ALL THEIR ENTHUSIASM FOR BADDECK, BOTH ALEC AND Mabel enjoyed visiting Europe frequently, and the girls even went to school abroad.

Their new house on Connecticut Avenue in Washington possessed the winding staircase Mabel had especially wanted, remembering her childhood delight in the one in the Cambridge house. The plumbing pipes installed outside the house because of Alec's fear of sewer gas gave the house a weird appearance and mystified sightseers for years.

When Elsie and Daisy gathered their friends into the friendly atmosphere of the house for parties and dances, the young people of Washington were just as enchanted with Mabel Bell as the young people of Baddeck had been. Gathered around the long table, they would thump on it to capture her attention, and she became the confidante of them all. She was still the "creative listener" she had been in her teens. Daisy commented, "I think it was because you could actually confide in her without making a sound—just by moving your lips—and tell her things it would have been hard to say out loud—and she always understood what you needed, and knew how to help."

Proudly showing off her peculiar immunity to dizziness, her daughters delighted in taking turns waltzing her around and around until both of them were too dizzy to stand up, whereupon their mother would calmly walk away.

It was the era when all fashionable dresses possessed innumerable hooks and eyes or rows of aggravatingly small buttons and buttonholes always completely beyond the wearer's reach, so when Elsie and Daisy were getting into —or out of—their party frocks they would seek out their father, perch on his knee, and ask him to do the honors. He always did, groaning all the while over the foolishness of the clothes his womenfolk chose to inflict upon themselves.

"You don't think much of men's clothes, either, Daddysan!" Daisy would tease him. (She had discovered the Japanese suffix "san" which meant "honorable," and had adopted "Daddysan" as her own special name for her father.) "You think they are just as stupid as ours!"

"So they are," her father declared. "Tight trousers, tight coats, tight shoes—I had too much of them when I lived in London with my grandfather!" Comfortable tweed knickers with a soft silk shirt and a Norfolk jacket had become his trademark; he wouldn't wear shoes that had to be laced; and he considered it "a senseless waste of time and effort to fuss with a necktie," so he wore the kind that was already tied.

Mabel generally left a pot of hot chocolate waiting for her daughters after a party, and she never discovered it wasn't wanted, because their father regularly drank it, strictly against his doctor's orders. He never paid any attention to any doctor's orders.

For all their Washington interests, the Bells' equal love would always be given to Baddeck and *Beinn Bhreagh,* and one of the greatest fascinations of their first years there was the houseboat Alec had acquired. As Elsie described it later,

The Mabel *of Beinn Bhreagh was a real personality in my childhood, and many were the happy cruises we made on her. It must have been quite an edifying sight to see us start off—first a tugboat or launch, then the house-*

221

boat, and trailing on behind, sailboats, rowboats and canoes!

John MacDermid would take horses and buckboards, and stay at a nearby farm. The houseboat was just a little wooden house on two hulls. It was a wonderful sight to the country people, and they used to speak of it reverently as "the Floating Palace," and we always had many visitors to marvel at the completeness of our floating home. Everyone from miles around would come to visit us.

They were always particularly interested in the square hole in the living-room floor that opened into the water beneath the hulls, because not only was it a useful place to empty floor sweepings, but a fine place to fish! My Grandfather Bell and Great-Uncle David used to draw up arm-chairs and fish in great style!

All one's ideas of fresh air have changed so much that it seems too absurd to say that Father slept down in a dark, windowless spot in the hull when I was a child. He couldn't bear the slightest bit of light on his eyes!"

Charles Thompson, cook, butler, valet, courier, etc., generally went with us on the Mabel, *and we would stay for several weeks. The tug or launch would bring mail and provisions twice a week, and guests would come and go.*

But *Beinn Bhreagh* was far from being a mere holiday spot. In Washington the teacher and scientist in Alexander Graham Bell dealt with the problems and work with the deaf. There, also, he made studies in sound, heredity and longevity; but Baddeck seemed the perfect place to work out the myriad inspirations in his brain. *Beinn Bhreagh* was not only the perfect "dreaming place," but with its cool beauty, its spaciousness, and the eagerness of the intelligent, loyal men in the village to become his assistants, for the first time in his life he could really lift and spread the wings of his imagination. Probably the most versatile scientist America ever knew, Bell exercised all his talents at *Beinn Bhreagh.*

He turned sheep-raiser to prove that it is possible to produce a breed of twin-bearing sheep, and accomplished it despite the skeptical chuckles of the local farmers. So enthralled was he with every detail of his flock that he had a "village" of tiny houses built for his sheep at the top of *Beinn Bhreagh,* called it Sheepside, and laid it out in miniature streets, each with its own name. The sloping terraces leading to the Bras d'Or made wonderful pastures, but his wife and daughters never quite shared his enthusiasm. Mabel reported a little wryly, "Nearly every day for some years, he conducted his sometimes reluctant family up the mountain, often through deep snow, to visit these sheep, and to become personally acquainted with General Grant and General Lee, the first sires of the breed!"

He had a keen interest in medical science and a fund of practical knowledge. When Daisy suffered a particularly violent form of whooping cough, and had frightening spasms of choking, her father's knowledge of the anatomy of the throat and breathing mechanism enabled him to lessen the paroxysms and give her more relief than the doctor's medications. And when young Douglas McCurdy nearly blew off his right hand with a home-made bomb, it was Alec Bell who took charge, and persuaded the doctors not to amputate.

Alec was distressed and disturbed when he heard or read accounts of shipwrecked sailors or fishermen lost in fogs who died for lack of fresh water. Why hadn't someone discovered a way of distilling the salt out of sea water, or of condensing water out of fog? He built a laboratory, invented his own apparatus, sought men from the village who would be interested and willing to be his assistants, and plunged into the problem. Solar stills, looking for all the world like beehives, appeared along the paths of the laboratory. His process worked. Then he evolved a method for producing moisture from human breath—not very palatable, it must be admitted, but drinkable enough to save a shipwrecked man's life.

Alec Bell welcomed and applauded inventions that would supersede his own. When the miracle of the X-Ray broke upon the scientific world in 1895, he was thrilled and excited, although it would obviously replace his own electrical surgical probe, and he promptly began to experiment with it. Two years later Mabel was writing to tell Daisy how he had used it to locate a needle imbedded in a man's foot. She herself had seen the slide with the location of the needle under the big toe clearly to be seen, and the doctors had decided to operate at the Bell laboratory.

Whereupon Papa worked himself and all of us into a great state of excitement, discussing the chances of the man's fainting or even dying on his hands! Mr. McCurdy said he had an engagement to go to Coffin's Island, and Papa said he would like to get away also. I was thankful I was going to Club and Board meeting that afternoon! As we were going to Club Mr. McCurdy telephoned that the operation had been performed and was a great success, and they were all greatly excited and pleased. At his suggestion we had Dr. McKeen come to the Club and show the X-Ray picture and explain all about it. Dr. McKeen said that the excitement and one of Papa's cigars had so affected his heart that he couldn't speak, but he did, and most interestingly!

This was the first medical use of the X-Ray in Canada.

The year of 1897 was an eventful year for the family. Besides excitement and pleasant triumph, sorrow came to them in December when Gardiner Greene Hubbard died at Twin Oaks.

It had been given to Gardiner Hubbard to pack more outstanding achievements into one lifetime than three or four other men would accomplish together: He had been an outstanding lawyer, practicing before the Supreme Court; he had established the first street car line between Boston and Cambridge and brought a pure water supply and gas lighting to Cambridge. He had fought to preserve his deaf child's speech and give her a normal education; and then had striven to win the same right for other deaf children when he campaigned for the Clarke School at

Northhampton, which he served as its president from 1867 to 1877 and as a trustee until the end of his life.

He had been practically the first and the best friend the telephone and its inventor had; he had served as Regent of the Smithsonian Institution and helped to organize the Church of the Covenant in Washington, now known as the National Presbyterian Church; and in January of 1888, he and thirty-two other men had started a society to interest Americans in geographical matters, the National Geographic Society, whose president he had been until his death.

The Society elected Alexander Graham Bell its president in January, 1898, and Gardiner Hubbard would have been pleased and grateful at the way his son-in-law led it into vigorous new fields. It had been a little like a select, scientific and rather local club, occasionally publishing a paper-covered pamphlet for its members called *The National Geographic Magazine,* which contained such enticing articles as "Geographic Methods in Geologic Investigation," or "The Classification of Geologic Forms by Genesis."

The new president had been enthusiastic about the Society since its inception, but he was not satisfied with the fact that eight hundred of its one thousand members were Washingtonians. He wanted the Society to grow and its magazine to be distributed all over the country. The thing he needed most was a creative editor, one with both literary knowledge and business ability, preferably a young, energetic man, with the imagination and talent to popularize geography, and make the magazine something to be noticed. But where and how to find him?

And then came one of the curious quirks of destiny with which the Bell story seems to abound. Alec received a telegram from Mabel:

> *Mrs. Grosvenor invites children to Amherst for Commencement Week. May they go?*
>
> *Dutiful Wife*

225

Two years before Professor Edwin Grosvenor had come twice from Amherst to lecture before the National Geographic Society. Afterwards he had been entertained by the Bells, and like all fathers, had launched into stories about his sons. They were identical twins and he was justifiably proud of them. Mabel listened fascinated. They sounded like wonderful boys to her. Perhaps she was thinking of what her own two sons might have been.

Several months after Professor Grosvenor had been their guest, she read an article in *Harper's Weekly* about the boys' graduation from Amherst. They had amazed and amused everyone as they outstripped and outranked everyone in their classes, accomplishing the same feat in athletics when they walked away with two championship tennis cups instead of one. They had crowned their college career by being twin valedictorians of their class.

Then Mabel had invited them to spend three weeks of their summer vacation at *Beinn Bhreagh,* where they had delighted everyone. Alec Bell smiled at the recollection. There was a promising romance in the making there, although for the life of him Alec wasn't too sure just which twin was involved. Elsie always knew, however. "Bert is the handsomer one!" she would say.

Chuckling, Alec sent a telegram of his own:

Children may go if they won't flirt too much.
                                        Dutiful Husband

It seemed to Alec that one of the Grosvenor twins might be the creative young man he was seeking, not only to gain subscribers for the *National Geographic,* as any editor might do, but to make the magazine represent something so outstanding that it would win members for the Society.

In February of 1899 he wrote a very frank letter to Professor Grosvenor, telling him he was on the lookout for a promising young man to become Assistant Secretary of the National Geographic Society and to manage the publication of the Society's magazine. The salary would be

226

low, one hundred dollars a month—he neglected to mention that he would pay it himself—but it might serve for an unmarried man. Would either of the Grosvenor sons be interested?

When the letter reached them Edwin had already decided upon a law career. Gilbert was teaching in a boys' school in New Jersey at a higher salary than Dr. Bell offered, but after a schedule that included teaching French, German, Latin, college algebra, chemistry, public speaking and debating, managing a small magazine must have seemed like an opportunity to draw a long, free breath once more. And there was another lure: Elsie Bell was in Washington, and if he were working for the Society headed by her father . . .

She was present when he went to Washington for an interview, and when she smiled at him and whispered, "I told Papa I thought *you* had the talents he was looking for, and would like to come to Washington," the matter was settled for him.

He promptly and enthusiastically agreed to become the first employee of the National Geographic Society on a day some people might have regarded as inauspicious—April 1, 1899.

Elsie was a good prophet. Gilbert had all the talents her father was looking for and more. Within five years he had created a *National Geographic Magazine* that became a byword for quality and distinction—a source of adventure and beauty. The magazine attracted so many new members that it became self-supporting in 1904, and Alec Bell could say to the young editor and director of the Society, "Bert, as you are now competent to paddle your own canoe, I wish to retire as president and devote all my time to my other interests—the deaf, eugenics, sheep-breeding and aviation."

Gilbert Grosvenor continued as editor, and became president in 1920. When he retired in 1954, the Society he had directed for fifty-five years had reached a membership of 2,150,000 and sponsored many scientific expeditions.

Far back in his childhood Alec Bell had talked about men voyaging through the air, and grew up with the conviction he would live to see flying machines. Moreover he had a deep-rooted premonition that he himself would invent a successful one. Ideas had beat against his brain while he was working on his first telephone, but he had resolutely shunted them aside. However, he had made Thomas Watson promise to set to work with him on the enterprise when the Bells returned from their English honeymoon. Two months after his marriage, on a walk along the English coast, he had pointed out to Mabel some gulls in flight, and she wrote, "ever since, his mind has been full of flying machines!"

Nothing came of the proposed Bell-Watson project, but at the beginning of the 1890's Alec's own ideas flared anew. Professor Samuel Langley, who was making serious experiments on the subject, was often at *Beinn Bhreagh* and word was spreading around about the "ridiculous" plans of Wilbur and Orville Wright. Certainly the great stretches of land, the open sky and the waters of the Bras d'Or, all combined to make *Beinn Bhreagh* an ideal testing ground. In 1893 Alec made a laughingstock of himself when he predicted to a reporter, "I have not the shadow of a doubt that the problem of aerial navigation will be solved within ten years!" But Alec Bell was used to being scorned. The Wrights flew exactly ten years after the interview.

Alec began his own experiments with ideas for "rocket propulsion," and spent many months testing everything to find a satisfactory fuel—from alcohol vapor to gunpowder. The models soared as high as two hundred feet into the air.

But before he would even attempt to design or build "flying machines" he decided to experiment with kites, trying to evolve structures which would be safe, strong and predictable in the conditions they would meet in the air. When he understood these, he reasoned, he would have a sound basis to design his "machines"; so he built a "kite house" on the estate and engaged whole crews to help him work out this dearest dream of his life.

They built all kinds of kites, all sizes—small kites and giant ones—of all varieties of materials. He tested his designs again and again, discarding one after the other, and finally coming to the conclusion that the tetrahedral design, with the kite's cells made up of four plane triangles, was the most satisfactory for all purposes.

Even the women of Baddeck worked on the Bell kites, covering their many "cells" with a light, thin, but very strong brilliant red silk, and the kites flaunted their queer shelves high over the hills and the lake, looking like strange, overgrown scarlet birds in the sky. The weirdest idea of all to the loyal crews was Dr. Bell's avowed intention of producing a man-carrying kite. They wouldn't have said so, but they probably agreed with the opinion of Dr. Bell's scientist friends, who were outspoken about their disappointment in him. Poor Dr. Bell! So unfortunate he had let this obsession with flying get the better of him! Even his old and loyal admirer, Sir William Thompson, now Lord Kelvin, wrote a letter of protest—not to Alec; he probably sensed it would fall on stony ground—but to Mabel. It was very evident that Lord Kelvin feared this obsession with the belief that man would fly in a guided machine would wreck Alexander Graham Bell's reputation as a responsible scientist once and for all. He spoke of "trying to dissuade him from giving his valuable time and resources to attempts which I still believe can only lead to disappointment, if carried on with any expectation of leading to a useful flying-machine."

Mabel had never in her life attempted to "dissuade" Alec from any of his inventions, and she wouldn't have begun now. She, too, was beginning to believe in human flight.

But regardless of what even his most valued colleagues might think, the Bell kites continued to ride into the Cape Breton sky. And everyone had a hand in maneuvering them. Mabel measured their pull against the wind. Guests were intrigued into taking a hand. Even Helen Keller helped, the summer she and Annie Sullivan spent her college vacation at *Beinn Bhreagh.*

" 'We are getting on swimmingly!' " she reported Dr. Bell as saying, and she added, "this was not infrequently true, for a recalcitrant breeze would catch us, and we would find ourselves swimming, not in the air, but in the Bras d'Or. Once when I was holding the cable, someone released the kite from its moorings, and I was nearly carried out to sea hanging on to it! Dr. Bell insisted that I should wear a helmet and a waterproof bathing suit, just as he did. 'You can never tell what perverse idea a kite may get into its head,' he would spell to me. 'We must always be ready to outwit it.' "

For well over ten years the kites flew high without the actual "flying machine" stage apparently drawing any closer, and then Mabel took a gentle hand. She had once said, "my husband would still be tinkering with the telephone if I hadn't taken it away from him!" Perhaps she thought the time had come to push him gently past the kite phase of his experiments. She managed it very adroitly.

Douglas McCurdy had grown up and entered Toronto University to study engineering. As he was leaving Baddeck after a vacation, Mabel spoke to him confidentially. "Now, Douglas, you will be coming back to us soon. We need you here. And if you ever run across another bright young man who might be able to help Dr. Bell in his flying experiences, bring him along with you!"

Douglas promised, but the thought slipped out of his mind and was gone until one June night in 1906 when he was packing his trunk to go home for the summer, and a friend casually wandered into his room.

"Casey," Douglas inquired on the spur of the moment, "Where are you going to spend the summer?"

Young Mr. Baldwin's name was actually Frederick, but

his passion for baseball had won him the Irish nickname. He shrugged in answer to Douglas' question. He had just graduated, but the faculty had a very low opinion of his future because it was known that he had wanted to do his thesis on aerodynamics, and he had spoken enthusiastically about the dawn of flying machines. In consequence there was no job on the horizon for the rash Mr. Baldwin, and his summer plans were as clear as a Halifax fog.

Douglas told him of Mrs. Bell's invitation. "Come to Baddeck with me. You'd get on famously with Dr. Bell!"

Casey had never heard of Baddeck, and he said so, but after Douglas's description of the place and particularly of the kite experiments, his interest was stirred. He decided to go for a couple of weeks and look the place over—and remained there for forty years.

He received the friendliest of welcomes at *Beinn Bhreagh,* and promptly lost his heart to his hostess. Dr. Bell he regarded with respectful awe.

In the following summer, after Douglas McCurdy had graduated, he found Casey and two other young men working on a plan to install a motor in the largest kite. Alec Bell had discovered Glenn H. Curtiss at a New York Motor Show. He was a motorcycle expert, and a builder of motors at Hammondsport, New York, and Alec had asked him to come to Baddeck to look into the possibility of motorizing kites. There was a fourth young man, a Lieutenant in the United States Army, Thomas Selfridge. The United States Army was beginning to think there might be something to this flying business, and had sent Lieutenant Selfridge to learn about the Bell kites.

So the four young men who were to become the pioneers of aviation, with Douglas McCurdy as the youngest at twenty-one and Glenn Curtiss the eldest at twenty-eight, came together and worked together through the summer.

The four of them, with Alec Bell, came into the great living room one wet, miserable, windy September night to gather around the comfort of the blazing fire and discuss the kite experiments of the day. Mabel was presiding over the tea tray, but she managed to watch the conversation very intently. Perhaps it struck her that this was the psychological moment to change the course of the aerial events, for she paused in her tea-pouring, looked over at her husband, and interrupted the technical talk, something very rare for her.

"Now, Alec, you have four pretty smart engineers here, and they are all just as interested in flight as you are, so why don't we form an organization for the purpose of building aircraft and getting a man in the air? I have just sold some property in Washington, and I will give you twenty thousand dollars for your work if you will promise to try to build something a man can really fly!"

✐ 22

A Man in the Air

FROM THE LUMINOUS LIGHT THAT FLASHED INTO HER HUS-
band's eyes Mabel knew her inspiration had succeeded.
But Alec Bell never leaped at any decision, and Mabel
understood when he set his teacup down, excused himself and strode upstairs. She smiled reassuringly at the
group. "Mr. Bell needs time to think the idea over," she
explained, "but I am confident of what he will say when
he returns."

She was right. Within an hour he was bounding downstairs, as full of excitement and eagerness as if he, too, were
in his twenties.

"Gentlemen, I think Mrs. Bell has made us an excellent
proposition, and I will be honored if you will join me in
such an association! If you are willing, let us discuss the
matter at once!"

So at Mabel Bell's inspiration the Aerial Experiment
Association was born on the night of September 30,
1907, before the fire in the *Beinn Bhreagh* living room, but
to make the matter legal and binding Alec Bell took the
four young men to Halifax the next day to have the compact they had evolved officially drawn up, signed by a
notary and witnessed by the Counsel-General of the United
States.

"*WHEREAS, the undersigned, Alexander Graham Bell,
of Washington, D.C., U.S.A., has for many years been*

*carrying on experiments regarding aerial locomotion at his
summer laboratory at Beinn Bhreagh, near Baddeck, N.S.,
Canada, and has reached the stage where he believes that
a practical aerodrome can be built on the tetrahedral principal, driven by an engine and carrying a man, and has felt
the advisability of securing expert assistance in pursuing
the experiments to their logical conclusion, he has called
to his aid Mr. G.H. Curtiss, of Hammondsport, N.Y., an
expert in motor construction, Mr. F.W. Baldwin and Mr.
J.A.D. McCurdy, of the Toronto Engineers, and Lieu-
tendant T. Selfridge, of 5th Field Artillery, U.S.A., military
expert in aerodromics, and*

*"WHEREAS, the above-named gentlemen have all of
them given considerable attention to the subject of aerial
locomotion, and have independent ideas of their own
which they wish to develop experimentally, and*

*"WHEREAS, it has been thought advisable that the un-
dersigned should work together as an association in which
all could have equal interest, the above named gentlemen
giving the benefit of their assistance in carrying out the
ideas of the said Alexander Graham Bell, the said Alexander Graham Bell giving his assistance to these gentlemen
in carrying out their own independent ideas relating to
aerial locomotion, and all working together, individually
and conjointly in pursuance of their common aim to 'get a
man in the air' by the construction of a practical aerodrome
driven by its own motive power and carrying a man."*

Some people chortled over "Mr. Bell's love of ceremony
and his dramatic methods," but remembering the Honorable George Brown's dealing with the telephone patents
he was supposed to offer to the English office, and the
twenty long years of heartbreaking litigation, Alec Bell
put a legal touch on everything he did, and who can
blame him?

Perhaps the most amazing thing about the compact was
its guarantee that each of the four young men was to be

considered equal to the famous Dr. Bell, and share in all
the honors which might come to the Association.

Alec Bell detested the word "aeroplane" just as heartily
as he disliked hearing "hello" over the telephone, but he
found himself alone on both counts.

Four of the signers of the compact were free agents,
but the United States Army still had its claim on Thomas
Selfridge, so Alec Bell went directly to the White House
and made his request for the Lieutenant's service to President Theodore Roosevelt. Besides being a farsighted and
pioneer-minded President, that gentleman knew an exciting
proposition when he heard one, and he promptly issued an
order giving Tom Selfridge indefinite leave to work with
the A.E.A.

Each man had his own role. Glenn Curtiss was Director of Experiments, Casey Baldwin was Chief Engineer,
Thomas Selfridge acted as Secretary, and Douglas McCurdy
was Treasurer, with Alexander Graham Bell as Chairman
of the Association.

All that summer everybody had been concentrating on
one massive kite, the *Cygnet*, planned to be the first to
carry a passenger, and the Associates decided that the
first business of the A.E.A. must be to finish and fly her.

Douglas McCurdy said that "she resembled a giant slice
of honeycomb, with each of her red cells being triangular."

There was no motor installed on her as yet, but on
December 6, 1908, all the conditions seemed so exactly right
for her flight that they unanimously agreed to test her,
and Thomas Selfridge was chosen for the honor of crawling
into the troughlike hole that served as a cockpit. He was
obliged to lie face downward. Of course there was no way
of steering the kite, but the idea was that when a gust of
wind struck, he would shift his weight so that the *Cygnet* returned to an even keel, and if the kite nosed downward he
was supposed to throw his body backward. When he had
wedged himself into his cramped cockpit he could see
nothing in any direction except for a few inches directly

in front of him, "and he was literally buried in hundreds of silk cells!" Nevertheless he made the flight. The A.E.A. had engaged a local steamer which cruised the Bras d'Or, the *Blue Hill*, to tow the kite into the wind, and when Alec Bell gave the signal, and the steamer started, the *Cygnet* did exactly what was expected of her, rising into the air with beautiful grace.

Having witnessed the consummation of all Alec's kite work, the thrilled Mabel wrote, "It soared immediately to a height of one hundred and sixty-eight feet in a twenty-five mile breeze, retaining an even keel through all the gusts, at times seeming as if glued to one spot in the sky. Finally, in a sudden lull it descended, alighting on the water so gently and evenly that its passenger was not aware of what was happening."

Unhappily he was only too soon aware. Enmeshed as he was in the flapping silk, Tom Selfridge hadn't noticed when the kite started its descent. A sailor aboard the *Blue Hill* had been charged to watch out for this emergency and stop the boat in time, but just as the *Cygnet* headed for the lake, a cloud of black smoke from the steamer's funnel hid her from sight. With the *Cygnet* unnoticed in the lake, the *Blue Hill* raced on, and shattered the kite before the bystanders' horrified eyes. Tom Selfridge struggled free of the wreck and was unharmed, but the greatest and the last of the Bell kites was demolished.

It was too late in the season to do more outdoor work in Baddeck, but Glenn Curtiss invited the A.E.A. to come to his motorcycle factory in Hammondsport, New York, and begin work on an actual flying-machine. The three young men went, but Mabel had a sudden and serious illness, and the Bells headed for Washington. However, the A.E.A.'s Chairman kept in close and constant touch with everything that was happening in Hammondsport with a weekly bulletin, and went up on inspection trips as often as he could. These "bulletins" finally totaled nearly two thousand pages of valuable data, and Dr. Bell's letters gave important suggestions.

What was happening was exciting. The five pioneers had drawn plans for a true flying-machine—*Drome Number 1*, they called it, and all of them were giving their full time and talents to building it. It was a glider, like the Wright brothers' plane, and it would be equipped with one of the Curtiss motors; but the wings, instead of being straight, as on the Wright glider, tapered to a point like a bird's wing. They built a little enclosed seat for the pilot, and because they intended to fly her from the ice-covered Lake Keuka, they provided her with sledge runners. When she was finished they called her *Selfridge's Red Wing*, because her wings were covered with the silk left over from the ill-fated *Cygnet*.

Watching all this Alec Bell made an almost uncanny prophecy:

> *I have no doubt but that in the future, heavier-than-air machines of great size, and of a different construction from anything yet conceived of, will be driven over the earth's surface at enormous velocity, hundreds of miles an hour, by new methods of propulsion. . . . Think of the enormous energy locked up in high explosives! What if we could control that energy and utilize it in projectile flight!*
>
> *We may conclude that neither our Army nor Navy could defend the United States from attack through the air. This requires the addition of a third armed force system of military defence, a National Air Force quite distinct from the Army and the Navy, capable of co-operating with both and also of acting independently of either. This might well be made a distinct department of the government, on the model of the Army and Navy Departments, and should be provided with a special college upon the model of those at West Point and Annapolis.*

It was the twelfth of March, 1908, and they trundled the *Red Wing* out upon the lake amid a crowd of spectators, and Casey Baldwin climbed into the cockpit. Someone spun the propeller and hurriedly stepped aside. The *Red Wing* glided over the ice at almost twenty miles an hour; then Casey pulled the lever and she rose—and flew! She wasn't too steady—she wobbled in a very disconcerting and uncomfortable manner—but she flew, and as they looked up at the red-winged glider sailing above their heads against the blueness of the winter sky, a cry of awed admiration sounded from the watching group:

"Gee, isn't she *pretty!*"

The marvelous flight measured all of three hundred and eighteen feet and eleven inches, but it had been a flight, and it had been a success. It was also the first public flight of a heavier-than-air craft in America. True, the Wrights had flown four years before, but their attempts had not been advertised. Alec Bell, on the other hand, was delighted to have witnesses. Several hundred awed spectators watched the *Red Wing* make her maiden flight and crowded wildly around Casey as he climbed out.

"What was it like? Wasn't it wonderful? Weren't you thrilled?"

Casey's calm was even more startling than his adventure. "Well—no. There wasn't anything to be excited about. I always knew she'd fly."

The precious machine came to grief less than a week later, but the A.E.A. was already working on Drome Number 2, *Baldwin's White Wing.*" She possessed movable attachments on her wing tips—"ailerons," a device Alec Bell had suggested to keep the craft stable. Everybody flew her, but it was Douglas McCurdy's ill luck to crash with her in a field scattered with tree stumps.

Drome Number 3, *Curtiss' June Bug,* was an improvement on them all, and when he flew her on July 4, 1908, he won the *Scientific American* trophy for the first measured official flight in the United States.

These flights by the four young men of the A.E.A., made in the machines they themselves created, not only turned

Alec Bell's lifelong beliefs into reality but woke the world to the realization that this was the dawn of man's era in the air. It was true the Wright brothers had flown to high altitudes and had made longer flights, but their tests had been done in such privacy there was no official proof. Mabel Bell's A.E.A. had brought aviation into the open for all to believe.

The most famous and successful of the "dromes," the one which really climaxed the Associates' work, McCurdy's Drome Number 4, the *Silver Dart*, was built and flown, like the others, at Hammondsport, but she made her most dramatic appearance at Baddeck, in February 1909. The most advanced craft of her day, she possessed stout spruce struts, taut linen wings, bamboo rods, the latest Curtiss engine, and a landing gear made of a tricycle arrangement of motorcycle wheels.

The Bells were proud of young Douglas McCurdy, of his growing success as a pilot, and of the *Silver Dart* itself. All Baddeck was invited to witness the flight, and many of the villagers were on hand the morning of the twenty-third, when the *Dart* came gliding over the ice, towed by the big red Bell sleigh, carrying Alec Bell, John MacDermid, and pretty Mabel McCurdy, Douglas's cousin.

The *Dart* was the first aircraft to appear in Canada, and it was an awesome sight. Those who had come on skates gathered about, touching it with mittened hands and dawning respect, still mixed with incredulity. Then Douglas appeared, held an inquiring finger to the wind, and climbed into the cockpit. One of the Bell crew spun the propeller, and the *Dart* lived up to her name as she shot forward.

Years later Douglas reported with obvious enjoyment, "*Then* the crowd screeched and scrambled away!" The *Dart* lifted itself, soared to a height of sixty feet, and flew steadily, evenly, gracefully, leaving the people of Baddeck gazing upward open-mouthed. It was a short flight, half a mile; then the *Dart* swung about in an easy turn, and skimmed back.

This time the skaters were waving and shouting, clamoring to reach the town hero, but Douglas was looking at Dr. Bell, delighted with his praise:

"A beautiful flight, my boy, a very beautiful flight!"

"Thank you sir, it was just a test. Now I'll really fly!"

But Alec Bell laid a firm hand on young Douglas's arm. "No, my boy, put the *Dart* away. There'll be no more flying today. Fly tomorrow, or the next day, but today is almost a sacred day. We'll have nothing to mar it!"

The day was a triumphal high point for the A.E.A., and Mabel wrote to Daisy, "The Town Council of Baddeck voted to enter the event on the town records and sent Papa and Douglas copies of the resolution. It was a beautiful flight Douglas made this morning toward the lighthouse, making a glorious sweep and back past B.B. shore, passing close to me purposely. The drome is such a beautiful sight it goes to Daddysan's heart. He is so proud of Douglas. All the Associated Press despatches are written by him."

There had been a time limit on the Association. It was supposed to have expired at the end of September 1908, but it seemed to Mabel it was making such glorious progress that she gave another ten thousand dollars to extend it to eighteen months.

In August 1909 Tom Selfridge was summoned to Washington to join the newly formed Aeronautical Board of the Army, and just a few weeks later he was delighted by Orville Wright's invitation to join him in test flights at Fort Myer, Virginia. That certainly would mark the beginning of a wonderful collaboration between the already famous Wright brothers and the A.E.A.!

Mabel Bell was delighted as well. She had grown especially fond of Tom Selfridge, and she planned a celebration dinner for the night of the first flight. And then came the message which left her white and shaken. There would be no celebration. The Wright plane had crashed. Orville

With his daughter Elsie, Dr. Bell revisits the old Bell home (1870-1880) in Brantford, Ontario, where he had his inspiration for the telephone.

Arthur McCurdy photographed Mabel Bell gazing at her new summer home, Beinn Bhreagh Hall, on a wintry day in 1894.

Photograph from Gilbert Grosvenor

"Comes a pause in the Day's occupations . . ." Gertrude, Lilian and Mabel Grosvenor with "Grampie Bell."

Photograph by Charles Martin

The houseboat Mabel of Beinn Bhreagh, *used at first by the family for pleasure, then as a retreat by Dr. Bell.*

Men who developed the telephone visited the National Geographic Society on the telephone's 40th anniversary, March 8, 1916. Thomas Watson stands back of Dr. Bell, Theodore Vail at his right. Son-in-law Gilbert Grosvenor is second from top at right.

Photograph Harris and Ewing

Photograph by Gilbert Grosvenor

This photograph of the Bells strolling on Beinn Bhreagh's garden walk was Mabel's favorite picture of herself and her husband.

Wright was seriously injured, and Thomas Selfridge became the first person to lose his life in an air wreck.

His death also wrecked the Association. Mabel said sadly, "The beautiful bond of companionship which we called the 'Aerial Experimental Association' was broken when Tom Selfridge sealed his devotion to our cause with his life."

Douglas McCurdy made a career of flying. Casey Baldwin returned to engineering; Glenn Curtiss to his motors, although he also flew. But the A.E.A. had lived up to the bright vision Mabel Bell had for it that September night in 1907. It had made its mark upon a new science, and it had made Mabel Bell the first woman patron of aviation.

It also proved Gilbert Grosvenor's point when he said, "Mrs. Bell was as great a woman as Dr. Bell was a man. Despite her deafness she was exceedingly well informed on scientific matters. And, although history hasn't paid much attention to the fact, her creative talents seemed scarcely less brilliant than Bell's himself."

~ 23

Footprints on the Steps

EIGHTEEN MONTHS AFTER HE HAD COME TO WASHINGTON young Gilbert Grosvenor sought a leave of absence from the National Geographic Society and betook himself to London to marry Elsie May Bell. Much later, he would observe, "She gave a magic touch to the Society when she consented to marry me!"

Three years after the Grosvenor marriage, the Society brought a true love to Daisy when Gilbert Grosvenor invited David Fairchild, botanist and "plant explorer," to lecture before the Society on his expedition to Baghdad. To his surprise the young editor found that the scientist was not a bearded, middle-aged man. After the lecture Gilbert escorted him to one of Dr. Bell's famous "Wednesday Evenings," where he sat "contentedly in a corner, and felt immediately at ease, as one does with any really great and simple character." It was there he met Daisy—he preferred to call her Marian—and when he learned by some chance remark from Elsie Grosvenor that her sister was not engaged, he "left with his head in a whirl."

In 1905, David and Marian were married. Always the pioneer, Mabel gave the bride an electric car, in which she and David made joyous excursions into the country, driving at the then incredible speed of twelve miles an hour.

The marriages brought Mabel and Alec Bell the gifts for which they had been yearning all their married life—

242

two sons of whom they could be more than proud—and both the Bells rejoiced in the young men who had come into their lives. It was Mabel who said, "I think no mother was ever more blessed than I in her daughters and sons-in-law!" but Alec would have emphatically agreed that few fathers could claim a happier relationship. There was an exceptionally deep and strong bond between him and his sons-in-law. David Fairchild said, "The mere thought of Mr. Bell stimulates and enheartens me. It was one of the great joys of my life that I had the privilege of knowing him so intimately." Gilbert Grosvenor never ceased to treasure the old roll-top desk Alec Bell had given him the first day he reported for work in the cramped quarters of half a room of the Corcoran Building—the other half being occupied by the American Forestry Association.

Experience with his father's two-volume history of Constantinople (1895), with its innovation of two hundred and thirty photoengravings, had made young Gilbert enthusiastic about the use of illustrative photographs, especially for such a magazine as the *National Geographic,* but at that time, as its young editor said ruefully, "every dollar had to be stretched like a balloon." Having no funds to pay for photographs, he invested well over a month's salary in a camera, and planned to provide many of his own illustrations. He decided to begin with the article which Dr. Bell was doing about his kites for a forthcoming issue of The Magazine. Why not picture the kites through every phase from their beginnings to their triumphal flights skyward? A series of dramatic photographs resulted, and Gilbert Grosvenor had fallen in love with an art that would be a lifelong interest.

Outstanding as he was to become as an editor and in the fields of geography and nature, he now set himself an important labor of love for which there would be little credit. He was gripped by a desire to make a photographic chronicle of the Bell's lives. Sensing that there were two remarkable people, their son-in-law set about capturing and preserving the daily activities, the special moments, and the moods and personalities of Alexander Graham and Mabel Bell.

A few people, some of the A.E.A. members for example, complained that "Bert and his camera are always underfoot," but the photographs became a full and valuable record of the experiments and devices wrought in the laboratory at *Beinn Bhreagh,* and provided fascinating insights into the Bell family life, especially when Alec and Mabel assumed their new roles as grandparents.

A son arrived in the Grosvenor household first of all. Elsie Grosvenor named him Melville Bell for the grandfather she adored. And then came the morning when Charles Thompson woke Alec with the arrival of an urgent telegram.

"Who has signed it?" asked the sleepy inventor.

"It is signed 'David,' sir."

"Very well, go ahead, read it," Alec was trying to summon his faculties, always a difficult process for him in the early morning.

"'Alexander Graham Bell Fairchild says good morning to Grandpapa. Signed, David.'"

This was the period when the inventor's days and nights were centered on the business of flying, and what followed has always been the delight of his children and grandchildren, for Alec Bell sat up in bed, looked earnestly at Charles, and demanded, "Can he fly?"

Eventually there were to be seven children in the Grosvenor household. After Melville Bell came Gertrude, Mabel, Lilian and Alexander Graham Bell, whose life was a brief span of five years. Mabel said of him, "He had such power for love in his little heart. I never saw such a little creature so full of love and gentleness, so little self-assertive, and yet there was no weakness to him." When the fourth little daughter arrived, Gilbert begged that she be named for her mother, but she had been born on March 3, her grandfather's birthday, and he cabled from Ceylon, asking that

244

she be given his name. Then Lina McCurdy said wistfully that everyone in the family had a namesake except herself, so the baby became Elsie Alexandra Caroline, which was promptly shortened to "Carol." And then came Gloria, named in thanksgiving for the arrival of a fifth daughter.

In addition to Sandie—or Graham, as he was occasionally called—Marian and David Fairchild welcomed two little girls, Barbara and Nancy Bell.

Both "Grampie" and "Gammie" Bell adored all the children. Alec knew his proudest and happiest moments when he held a baby grandson on the back of his horse Champ, or was teaching his granddaughters to swim. And the children knew they would always be welcome at his laboratory. "Never turn a child away," he cautioned his assistants, "and never ignore their questions—they may be observing something we have overlooked."

Full of ideas to demonstrate the value and uses of his favorite tetrahedral design, Alec Bell had commissioned Casey Baldwin to build a tetrahedral observation tower at the top of Beinn Bhreagh Mountain to prove that in spite of seeming frailty a tetrahedral structure had the strength to support heavy weights and would defy any storm. Casey finished it in August of 1907, and Grandfather Bell proudly decided that five-year-old Melville was the proper person to unveil and dedicate it. Mabel and Casey tried to help the child, but in addition to the fact that he had stage fright, the sight of the eighty-foot tower must have been rather appalling to the little chap, "although I finally did what was expected of me!"

But there were other experiences with his grandparents that were pure fun. Sometimes he went to the movies with them. Mabel especially enjoyed the movies, particularly Westerns. The movies were silent affairs in those days, and she would be amused by the fact that the actors were not saying the words attributed to them in the legends accompanying the pictures. When they went to the theater, Alec would keep her informed of the dialogue, and it always fascinated the boy to see him "talking to her continually without making a sound!"

When the families were in Washington it became Melville's assignment to join his grandfather at his Georgetown office and walk home with him to Connecticut Avenue. Alec needed the walk, but he hated to take it alone, and Melville loved scampering along beside him. Very often on the way they would pause before a famous bakery with a tempting array of beautiful pastries in the window. Lured by the sight, and with their mouths watering from the delicious aroma which had reached them almost a block away, the man and boy would move more and more at a snail's pace as they approached the display of goodies, and Melville would wait anxiously for the twinkle in his grandfather's eye and his low chuckle as he queried, "Don't you think we should go in and have one?"

They both knew that "Grampie Bell" was on a strict diet, but "in we would march and return with a piece of delicious pie, or cake or dumpling."

On their walks they would talk about *Robinson Crusoe*, with marvelous additions to the original, and when they reached Baddeck that summer, they decided to live the story. They would do exactly as Mr. Crusoe had done—provide their own clothes, make their own fires and find their own food! They made the concession of wearing bathing suits, and set off enthusiastically enough: The first practical thing would be to build a fire. However, the ground in Cape Breton Island is generally damp, and all the wood seemed to be green. They had scorned to bring a knife, so they couldn't break the branches from the trees. They rubbed sticks and more sticks together until their hands were raw, but the fire stubbornly refused to appear. "Finally," said Dr. Melville Grosvenor, "we got hungry, and headed back to the houseboat, where we cooked ourselves a delicious meal of baked beans!"

But being with his grandparents wasn't simply a holiday for Melville. His grandfather had a very poor opinion of

246

the methods used for teaching arithmetic. In that day if a teacher asked, "What is two times four?" the pupil was supposed to give a detailed answer: "Two times four equals eight."

"Utterly ridiculous," said Alec Bell. "Say eight! The rest of it just burdens your mind!"

He gave the boy a sound foundation in mathematics and science, and his grandmother focused her attention on his reading and composition. His grandfather's secretary read Latin with him, and Casey gave him lessons in navigation.

One awful summer someone managed to convince Mabel Bell that ice cream was bad for growing children; so in place of the delicious Sunday homemade ice cream that the Grosvenor children had looked forward to all week, the floating island of Mabel's childhood appeared on the Bell table—that, or plain custard. Their father saved the day by secretly promising them ice cream if they would only eat the custard or the floating island as happily as possible.

The grandparents used psychology with some skill. Sandie had been impertinent to a member of the *Beinn Bhreagh* staff, and refused to apologize to her. "Very well," said his grandfather, "we will hold a court trial." He briskly appointed Mabel and John MacDermid as the jury, and summoned the offended maid to appear as the complainant, while he acted as the judge. Graham Fairchild was found guilty as charged and sentenced to make an abject apology —or else. The boy chose the alternative and John MacDermid was sent to find "a nice limber birch rod." He brought it back with great relish, but by the time the switch was half peeled, Graham decided to apologize.

Once Graham and his cousin Lilian, impressed by the surprising number of garter snakes at *Beinn Bhreagh* one summer, decided to make a snakeskin rug for Gammie! The supply of snakes ran out, greatly to Mabel's relief, but she had smiled on the idea, knowing it was a labor of love. "She was," wrote Dr. Graham Bell Fairchild, "very much the center of Baddeck, the source of all comfort and the instigator of countless picnics, expeditions, parties and amateur theatricals. Somehow she managed to keep everyone doing something and being enthusiastic about it without ever appearing to be running things. Baddeck to me meant perfect freedom, and yet much of what I did was, I know, suggested by Gammie."

Lilian felt very important to be chosen to "listen" at the phone for Gammie, and tried to remember the proper "Hoy, hoy!" for Grampie.

Small Carol, spending Christmas at *Beinn Bhreagh,* never forgot the sight of her grandfather coming back from Boston, looking exactly like Santa Claus and laden down with snow and mysterious packages.

Gertrude was responsible for the most lasting impression the grandchildren left at *Beinn Bhreagh.* Workmen were busy one morning laying new steps each side of the wide porch, and Gertrude gleefully walked all over the wet concrete. A wrathful workman bore her to her grandmother, suggesting pointedly that she be justly punished, but the idea caught her grandmother's fancy. She laughed outright, hugged the surprised child close, and then gave the workman orders that utterly amazed him: "Keep the concrete wet! I am going to get all my grandchildren together, and we will make footprints of *all* them on the steps!"

"But," Dr. Fairchild remembers, "all of us knew Grampie came first. There was never any question about that. The one iron rule was no noise in the mornings when he was asleep, and although Gammie couldn't hear us, there were plenty of loyal servants who could!"

The enthusiastic grandfather was still a busy scientist, at work on countless projects. One of them was distilling salt out of sea water in an apparatus so small it could be carried in an ordinary teakettle; another was pontoons for aircraft. Bell developed the first seaplane.

He disliked writing anything for himself, and from 1877 when he acquired his first secretary, he had made "dictations" a habit, and his "Home Notes" filled a great many volumes. They covered everything—his progress in breeding his twin-bearing sheep, his views on the President's politics, his serious scientific projects, and a granddaughter's fall out of a tree! They are signed by his secretaries, by Mabel, by Casey Baldwin—even, as they grew old enough, by one or two of his grandchildren.

Perhaps one reason he had so much time to enjoy his grandchildren was that he still considered the middle of the night the fittest time for serious study. This could not always have been especially convenient for Mabel, for if Alec was seized by a sudden and brilliant inspiration at three o'clock in the morning, he would wake her and tell her of it, and ask her to take his dictation for its development. Her daughters smilingly maintain that their mother wouldn't have foregone sharing these middle-of-the-night inspirations for the world!

One morning when Mabel looked at the gaping holes in her prized Venetian blinds, just received by a special order from Halifax, she could only stare in consternation. In the middle of the night Alec had had a sudden idea for a new type of propeller and needed some slender strips of wood. Perhaps it was on that day Mabel confided to a friend that being the wife of a great inventor was not always an easy life! The propeller is an important exhibit in the Alexander Graham Bell Museum at Baddeck.

Sometime later Alec was working on a project in his laboratory which required rapid evaporation. Neither he nor his assistant, John MacNeil, could think of the right insulating material. "Well," said Dr. Bell comfortably, at the close of the afternoon, "if we haven't come up with something by tomorrow morning, there are plenty of carpets at *Beinn Bhreagh* we can cut up. I think carpets would do the trick."

Remembering Mrs. Bell's face when she saw the Venetian blind, Mr. MacNeill resolved to think of some more suitable insulating material than the *Beinn Bhreagh* carpets—and did.

Bell never lost his innate kindliness, however. John Mac-Neil approached the inventor one day and respectfully asked if he might change some of the specifications of something he was working on. He never forgot Bell's wonderful smile and his quick answer. "Mr. MacNeil, will you ever get it through your head that you are working *with* me—and not *for* me?"

The inventor never approached the laboratory door without first knocking and inquiring, "Gentlemen, may I come in?" and he always received visitors—even unexpected strangers—with instant and genuine friendliness.

One day one of the Bell neighbors, Percy Blanchard, noticed four girls, all strangers, gazing wistfully across the lake. They told him shyly that they were "hello girls" from Boston, and had come to Baddeck on their vacations because they knew Dr. Bell lived there. Mr. Blanchard offered to take them in his boat to a spot where they would be much nearer *Beinn Bhreagh* than they could get by land. Arriving at *Beinn Bhreagh,* he escorted the awed girls to the little tetrahedral-shaped building from which Dr. Bell watched many of his kite flights. Mr. Blanchard said "he couldn't have treated them more magnificently if they had been princesses of the realm." He staged a very special exhibition of kite-flying, and then gave each girl a small kite-section with red silk wings as a souvenir.

He could be upset to the point of following trigger-quick impulses. Once he had designed and worked over a particularly huge kite which was his especial joy, but which only an especially strong wind could fly. He had waited impatiently several weeks for winds of the proper force. At last a howling gale swept over the hills and the bay, and the inventor rushed gleefully to the kite-house, but alas, the Bras d'Or was in such a state of lashing waves that the members of the kite crew had evidently decided that discretion was the better part of valor, and remained at home! Mabel wrote the results to Daisy:

> *Just imagine poor Daddysan's feelings when his big storm came, and not only one but all of his men absent because of it. He came home deathly white and went to his study and wrote a curt note of dismissal of everyone on his staff except MacNeil, and then went and lay down with his face to the wall without speaking or moving for hours!—But he took them all back the next day!*

When he worked at night he would go about smothering all the clocks with towels to make the house as soundless as possible. One morning his daughters laughed themselves weak when they found, well-blanketed, a certain clock that hadn't run for a year. Alec would work until near morning, then wrap his head in towels to guard his sensitive eyes against the morning light and fall into a deep sleep that would last for hours. If he was wakened before he was ready, "He would look just awful," according to his daughter, Elsie. "If that happened, he would have a terrible sick headache, and his hair, which usually stood up straight like a lion's mane, would fall lank and damp!"

Elsie could wake him with a gentleness that prevented the headaches, so before her marriage she was chosen to accompany her father on his out-of-town journeys.

As a Regent of the Smithsonian, Alec had once been delegated to go to Genoa, where James Smithson had died and been buried, and bring back his remains for an honored burial in Washington. The morning of the ceremony Charles Thompson tried to wake him for the event, but with slight success.

"Wake up, sir! Mr. Bell, you have to be at the Navy Yard at ten o'clock this morning for Mr. Smithson's funeral!"

Alec half opened his eyes and spoke impatiently. "What nonsense are you talking, Charles? James Smithson's been dead fifty years! Why has he come back to bother me?"

For several years, winter or summer, whenever he was

251

in Baddeck, he would pack a small satchel with papers or plans which he particularly wanted to study, and retreat for the weekend to the *Mabel of Beinn Bhreagh,* now retired from lake duty and drawn up on the shore, and there he would be alone except for the sounds of the woods and the soft lap-lap of the water on the pebbly beach. No one except Mabel was allowed to telephone him; and except for John MacDermid, no one was allowed to approach the houseboat. When two inquisitive boys came skating up the lake one day and decided to explore "the funny old house," they were startled by the apparition of a white-bearded man in a long red bathrobe who greeted them with "Good morning, gentlemen, what may I do for you? Will you share my coffee?"

On exceptionally rare occasions Mabel joined Alec on the houseboat for a Sunday, enjoying her "camping-out." It couldn't have been a very rugged experience, for John MacDermid spread rugs and blankets, Alec tended the fire, and, as Mabel said, "Kathleen sent in grapefruit and hot buttered toast, and all I had to do was to drop eggs into the crispiest bacon you ever tasted to have a breakfast fit for a king!"

Kathleen was the wife of Casey Baldwin, and must have been an especially delightful person, for Cousin Mary Blatchford wrote about the Baldwins, "Each one is nicer than the other, and I would have adopted them if Mabel hadn't seen them first and gotten ahead of me!"

✍ 24

Beautiful Mountain

CASEY AND KATHLEEN BALDWIN SHARED THE BELLS' JOURNEY
around the world in 1910. Experienced traveler though she
had been since her childhood, Mabel never in her life lost
her capacity for joy in seeing what Elizabeth Barrett
Browning called "pleasures new," and this trip opened new
worlds for both her and Alec. They had never seen India,
and Mabel, at least, had never beheld New Zealand or
Australia, and her letters to her daughters reflect her delighted absorption in every new sight and discovery.

In China the miniature streets and houses caught her
attention. "The narrowest of European streets are four
times as wide as Chinese streets," she told the Grosvenors,
"and the only means of conveyance are the sedan-chairs.
Across the canal the houses resemble wooden bird-cages, set
side by side one above the other. The little packing boxes
of the Chinese are chuck-full of the most exquisite things—
the shoes, the children's playthings, the cooking utensils of
the very poor are works of beauty and minute art."

The Bells and the Baldwins were invited to a lunch at
the "New Wa-Wu-Pu Building," the new Foreign Office in
Peking, and were impressed by the menu, printed in Chinese on one side and English on the other, and bearing
the Imperial crest. Mabel was enchanted with the strange
Chinese food: Chicken soup with pigeon eggs; fried silver
fish; shark's fin with sauce; mushrooms, Yunnan ham,

bamboo sprouts; bamboo sprout with caviar; pheasant; roast duck; fried shrimp; cakes with almond sauce; fruit and coffee. "I went through every item straight, and it was delicious! Papa said it was 'fine,' and came home and got rid of it!"

But it was really the "doll-like babies" that Mabel confessed dragged at her heartstrings.

In India she could laugh over "poor Papa's difficulties in getting a bath in the tiny tin tubs." And in Allahabad, India, the whole party had a thrill when they sent off their first airmail letters in a tiny, rather rickety plane at a carnival.

The scenery in New Zealand and Australia fascinated Mabel, as well as the odd brand of English spoken; and in Australia she met many of the Bell relations for the first time. Alec was overjoyed when his cousin presented him with a full-length self-portrait of his mother.

It was in Swiss Lake that the two men had their day. They visited and watched the Italian engineer, Enrico Forlanini, display his "hydrosurface" boat that roared through the water at forty-five miles an hour. When Signor Forlanini hospitably insisted that they take a demonstration ride, Alec hesitated more than a little, but was too courteous to refuse.

As far back as 1906 Alec had noted in the "Home Notes," "We are in the motor boat stage, aerialdromes in water become hydrodromes—why not tetrahedral winged cells in water to bring boat out of water?"

Both Alec and Casey were impressed with the Forlanini boat, but they thought they could do still better.

When they reached Paris both men grew excited over a seventy-five horsepower motor built by the Gnome Company. Alec thought it would be exactly the right thing to power the new man-carrying kite he was planning, the *Cygnet II*. When the Gnome Company refused to sell the motor unless one of their factory-trained men was given the privilege of installing it, Casey asked to enroll for a six-weeks special course, and he won the right to take the motor to Canada and install it.

The *Cygnet II* never flew, but the Gnome motor was an immediate success in the new type of boat that Bell and Baldwin were creating. They named them "hydrodromes" and then shortened the cumbersome word to "H.D."

There were four H.D.'s from 1910 to 1919, and the *H.D.-4* was a triumph for her creators. Used as the people of Baddeck were to the fantastic contraptions that emerged from the Bell laboratory, they must have blinked several times at the sight and sound of this weird object like an overgrown gray cigar rising headlong out of the water, lifting itself on what appeared to be metal stilts, and rushing across the Bras d'Or at the then incredible speed of seventy-one miles an hour. She was the fastest boat in the world for many years.

The craft leapt free of the water, leaving only "the smallest of foils" submerged, with the whole hull free of any water-drag, but experts from the British Admiralty and the United States Navy failed to see its potentialities. They came, they saw, they said it was "a surprising and marvelous invention," and they departed. Forty years later all the navies of the world would be developing a "new idea," the hydrofoil boat.

Mabel's interests, like her husband's, were taking her into new fields. She was becoming increasingly interested in women's battle for the right to vote, and she told her daughters one day, "I'm sending my motor-car in the procession tomorrow which will carry the petition to Congress. . . . I've also been to one of the meetings, and my pride is thoroughly aroused! The privileges accorded to persons of inferior education and position simply because they wear trousers and I don't! I don't particularly want to vote about anything myself, but I do object to being told I can't!"

Mabel had not been at all impressed by the kindergartens her grandchildren were attending, and what the famous editor Mr. S. S. McClure told her of Madame Montessori's school in Rome intrigued her. When she heard there was a Montessori class being held in one of the estates on the Hudson, she promptly went to investigate. She came away brimming with enthusiasm and determined that her grandchildren should go to a Montessori class.

She engaged two teachers and opened her own home to the school, but there were obvious difficulties in the way, so she found a house on Kalorama Road in Washington, and established the school there. She also organized one for Baddeck. "She was the godmother of both schools," David Fairchild remarked, "and as long as she lived she followed the careers of the little 'Montessaurs.'"

When the sinister cloud of the First World War loomed over the United States and Canada, everything was shunted out of the way except the fierce effort to "win the war to make the world safe for democracy." Baddeck people built boats, and Mabel Bell installed shelves for drying vegetables. She gave luncheons and dinners with menus made up entirely of dried foods, and reported that her dried spinach was especially popular.

Two pleasant occasions brightened the war years. On January 15, 1915, Alexander Graham Bell was invited by Theodore Vail, President of the American Telephone and Telegraph Company, to join with Thomas Watson in the first telephone call to span the United States. His original telephone was connected in New York City, and Thomas Watson waited in San Francisco. The telephone officials had written several appropriate messages for Dr. Bell to use, but he waved them all aside. He had decided upon his own, he informed them, and with everyone crowding breathlessly around, the father of the telephone took his place and waited for the signal. When it came he raised his voice: "Hoy, hoy, Mr. Watson! Mr. Watson, come here, I want you!"

And from over three thousand miles away came Thomas Watson's answering chuckle. "I'd like to, Dr. Bell, but this time it would take me a week!"

Melville Grosvenor was there with his grandfather, and now Alec moved aside, and motioned the boy to speak, standing by proudly while his grandson became the first child to speak across the United States.

There was another day to remember, in 1916, when Brantford took a day out of her somber, war-clouded life to honor the eager young man who had had the inspiration for the world's most important invention by the shores of the Grand River. Mabel and Alec stood hand in hand as an impressive monument to commemorate the telephone was unveiled and the old Bell homestead at Tutelo Heights was dedicated as a public park and museum by the Governor-General of Canada, the Duke of Devonshire. It is now maintained by the Telephone Pioneers of America.

Alexander Graham Bell would not have been human if he hadn't been pleased. "There are some things worth living for," he declared, "and this is one of them! When I came to Brantford in 1870, I thought I had six months to live. I am glad I survived to do this work and to witness this day!"

In 1921 there was a final trip for the Bells, to Scotland, this time, accompanied by the teen-age Mabel Grosvenor and Alec's young secretary, Catherine Mackenzie. Once again Alec visited and pointed out all the beloved spots of his boyhood. But they had not come to be mere sightseers. Edinburgh was about to honor her distinguished son. Mabel, almost hidden by her enormous bouquet of carnations tied with white and black ribbons, the colors of the city, proudly watched the pompous medieval ceremony in which the Lord Provost presented her husband with a parchment scroll proclaiming him a Burgess and declaring that he had The Freedom of the City. "I could only sit and be impressed and very pleased that Edinburgh had at last wakened to the realization that it had another son besides Sir Walter Scott to do honor to, and that it had not come too late."

As an "old boy" of the Royal High School, Alexander

Graham Bell had the privilege of giving the school a special holiday, and was delighted at the roar of thanks!

The scroll and its silver container were so precious in his sight that Alec Bell refused to have them packed, but carried them on the rest of their journey, and when customs officials would inquire suspiciously about the contents of the oblong box, he would inform them proudly, "This, sir, is The Freedom of my native City."

Four patents were granted to the Bell-Baldwin team for the "hydrodrome, hydroaeroplane and the like" in March, 1922, exactly forty-six years after Alec had received the first one for the telephone. These were the last.

Alec worked almost the same as usual, examining his sheep and busy with his reports, but something had changed. For years he had been under the shadow of diabetes, which he had gallantly tried to ignore, but the time had come when he could ignore it no longer. Ironically enough, if he could have lived another year, insulin would have come to his aid, but now there were a few days of mercifully brief illness, and then he slipped away from the circle who watched over him on the second of August, 1922.

Mabel planned the last thing she could do for him with loving detail. He lay in his comfortable tweed working suit, with the button of his prized French Legion of Honor in his buttonhole, and then his loyal crew carried him up to the top of his beloved mountain for the last time, accompanied by the sound of joyful music. Mabel stressed the importance of its being joyful, to remind his grandchildren that his life had been a joyous one.

In the next weeks Mabel poured her heart into letters she wrote to her grandchildren. She reminded Graham Fairchild that they were all depending on him to carry on the Alexander Graham Bell tradition, and to little Carol Grosvenor, who was spending a year with her Grosvenor grandparents, and who was heartbroken over the loss of her Grampie, she wrote one of the tenderest letters of all:

My Darling Carol:

Nannie writes me that you are feeling very badly because you have lost Grampie. I am glad that you feel sorry, because he loved you very dearly, and it shows that you, too, loved him very much.

Just a short time before he died he sent for Miss Mackenzie and gave her one of his "dictations." In it he said that no people ever had finer grandchildren than he and I. Wasn't that nice?

I miss my little Carol very much. It would have been lovely if all of you could all have been together with Grampie that last day. But it was too far for any of you to come, and anyway I am not sorry that you should remember him just as he was, happy in his grandchildren, his home—everything that he was doing.

And now don't forget, he still lives in you and all his grandchildren. Don't forget that you represent him now. It isn't necessary that you be great or famous—just be good and sweet, and thoughtful of others, and you will be like him, and carrying on his work.

Your Loving Grandmamma

Mabel's grandchildren always remembered her as a source of comfort, strength and joy. They remembered a person with the fragrance of lavender about her, her enviable skill with flowers, her gentleness and instinctive love of justice and fair play, and her joyous readiness for everything. "Listening" at the telephone for her, or holding her hand to steady her as she walked in the dusk or dark—such things were a real honor because they were done for "Gammie."

Only once, near the end of her life, did Mabel give anyone a glimpse into the lifelong struggle behind her. Then she confessed to her niece, Helen Bell: "all my life I have tried my hardest to have you children and everyone else forget that I am not the same as your mother, for instance."

Mabel Bell had run the course her mother had charted

for her the day she spoke the words, "*And goodness and mercy shall follow me . . .*"

The lives of Mabel Hubbard and Alexander Graham Bell had been so closely intertwined that one seemed only half alive apart from the other. Did Mabel have an inkling that she would not have to wait long to rejoin her husband? It was a very brief time, only five months, before she went to lie beside him at the summit of their "Beautiful Mountain."

The man who lies there was a scientist, prophet, humanitarian and teacher of the deaf, but the tablet set in the rock says simply, "Alexander Graham Bell, inventor, born March 3, 1847, died a Citizen of the United States, August 2, 1922."

Perhaps his own words, which greet a visitor as he enters the museum at Baddeck, describe him best of all:

> *An inventor is a man who looks around upon the world, and is not content with things as they are; he wants to improve whatever he sees; he wants to benefit the world; he is haunted by an idea; the spirit of invention possesses him, seeking materialization.*

Beside him is "his beloved wife Mabel Hubbard Bell, born November 25, 1857, Cambridge, Massachusetts, died January 3, 1923," a girl and woman whose days were wrapped in silence, yet who made both their lives a joyful sound.

POSTSCRIPT

I appreciate this opportunity to thank Macrae Smith Company and Miss Helen E. Waite for the beautiful book she has written about my parents, Mabel Hubbard Bell and Alexander Graham Bell. Miss Waite has caught the spirit of their lives with remarkable accuracy and charm.

ELSIE MAY BELL GROSVENOR

August 15, 1961
Beinn Bhreagh
Baddeck, Nova Scotia
Canada

APPENDIX

The documents reproduced on the following pages include Elisha Gray's letter giving Alexander Graham Bell full credit for the talking feature of the telephone, a remarkably prophetic letter from the young inventor written in London in 1878 and Bell's carefully detailed rebuttal to the Attorney General of the United States.

ANSON STAGER Pres!
ENOS M. BARTON, Sec'y

GEORGE H. BLISS Gen.Ag't,
ELISHA GRAY, Electrician

Office of
Western Electric Manufacturing Co
220 to 232 Kinzie Street.   Chicago Mar 5 1877

Prof- Bell
    My Dear Sir.
            I have just
rec.ed yours of the 2nd inst—
and I truly forgive you for any
feeling your telegram had aroused.
I found the article I suppose you
referred to in the personal column
of the "Tribune" and am free to say
it does you injustice.
I gave you full credit for the
talking feature of the telephone, as
you may have seen in the
associated press dispatch that was
sent to all the papers in the country—
in my lecture in McCormick Hall
Feb. 27th.   There were four different

papers represented at the lecture but only one—the tribune—alluded to any mention of you. Except the "press" dispatch. I described your apparatus at length by diagram.

Of course you have had no means of knowing what I had done in the matter of transmitting vocal sounds. When however you see the specifications you will see that the fundamental principles are contained therein. I do not however claim even the credit of inventing it, as I do not believe a mere description of an idea that has never been reduced to practice—in the strict sense of that phrase—should be dignified with the name invention.

Yours very truly
Elisha Gray

To the Capitalists of the Kensington, March 25ᵗʰ 1878
Electric Telephone Company

Gentlemen,

It has been suggested that at this our
first meeting I should lay before you a few ideas
concerning the future of The Electric Telephone together
with any suggestions that occur to me in regard to
the best mode of introducing the instrument to
the public.

The Telephone may be briefly described as
an electrical contrivance for reproducing in distant
places the tones and articulations of a Speaker's voice
so that conversation can be carried on by word
of mouth between persons in different rooms,
in different streets or in different towns.

The great advantage it possesses over every other form
of Electrical apparatus consists in the fact that it
requires no skill to operate the instrument. All
other telegraphic machines produce signals which require to be translated by experts and such instruments
are therefore extremely limited in their applications
but the Telephone actually speaks and for this reason it can be utilized for nearly every purpose
for which speech is employed.

The chief obstacle to the universal use of
electricity as a means of communication between
distant points has been the skill required to operate
Telegraphic instruments. The invention of

Automatic Printing Telegraphs Dial Instruments &c. has materially reduced the amount of skill required but has introduced a new element of difficulty in the shape of increased expense. Simplicity of operation has been obtained by complication of the parts of the machine — so that such instruments are much more expensive than those usually employed by skilled electricians. The simple and inexpensive nature of the Telephone on the other hand renders it possible to connect every mans house or manufactory with a Central Station so as to give him the benefit of direct Telephonic Communication with his neighbors at a cost not greater than that incurred for gas or water.

At the present time we have a perfect network of gas pipes and water pipes throughout our larger cities We have main pipes laid under the streets communicating by side pipes with the various dwellings enabling the inmates to draw their supplies of gas and water from a common source.

In a similar manner, it is conceivable that Cables of Telephonic wires could be laid under ground or suspended over head communicating by branch wires with private dwellings counting houses shops manufactories &c. &c. uniting them through the main cable with a central office where the wires could be connected together as desired establishing direct Communication between any two places in

the City. Such a plan as this though im-
practicable at the present moment will, I
firmly believe, be the outcome of the introduction of
the Telephone to the public. Not only so but I
believe that in the future wires will unite the head
offices of Telephone Companies in different Cities and
a man in one part of the Country may Communi-
cate by word of mouth with another in a distant
place.   I am aware that such ideas may
appear to you Utopian and out of place for we
are met together for the purpose of discussing not the
future of the Telephone but its present. Believing
however as I do that such a scheme will be the
ultimate result of the introduction of the Telephone
to the public I would impress upon you all the advisability of keeping this end in view that all present
arrangements of the Telephone may eventually be
utilized in this grand system.

I would therefore suggest that in introducing
the Telephone to the Public at the present time you
should on no account allow the control of the Conducting Wires to pass out of your hands.—
The plan usually pursued in regard to private Telegraphs
is to lease such Telegraph lines to private individuals
or to Companies at a fixed annual rental.
This plan should be adopted by you but instead of
erecting a Line directly from one point to another
I would advise you to bring the Wires from the

two points to the Office of the Company and there connect them together. If this plan be followed a large number of Wires would soon be centred at the Telephone Office where they would be easily accessible for testing purposes. In places remote from the Office of the Company simple testing boxes could be erected for the Telephone wires of that Neighborhood, and these testing places could at any time be converted into central offices when the Lessees of the Telephone Wires desire inter-communication.

In regard to other present uses for the Telephone the Instrument can be supplied so cheaply as to compete upon favorable terms with speaking tubes bells and annunciators as a means of Communication between different parts of a House.

This seems to be a very valuable application of the Telephone not only on account of the large number of Telephones that would be wanted but because it would lead eventually to the plan of intercommunication referred to above. I would therefore recommend that special arrangements should be made for the introduction of the Telephone into Hotels and private buildings in place of the speaking tubes & annunciators at present employed.

Telephones sold for this purpose could be stamped or numbered in such a way as to distinguish them from those employed for business purposes and an Agreement could be signed by the purchaser that the Telephones should become forfeited to the

Company if used for other purposes than those specified in the agreement.

It is probable that such a use of the Telephone would speedily become popular and that as the public became accustomed to the telephone in their house they would recognize the advantages of a system of inter communication. When this time arrives I would advise the Company to place Telephones free of charge for a specified period in a few of the principle shops so as to offer to those householders who work with the Central Office the additional advantages of oral communication with their trades people. The central Office system once inaugurated in this manner would inevitably grow to enormous proportions for those shop keepers would naturally obtain the custom of such householders. Other shop keepers would thus be induced to employ the Telephone and as such connections with the central Office increased in number so would the advantage to householders become more apparent and the number of Subscribers be increased.

Should this plan ever be adopted the Company should employ a man in each central Office for the purpose of connecting the Wires as directed. A fixed annual rental could be charged for the use of the Wires or a toll could be levied. As all connections would necessarily be made at the central Office it would be easy to note the time during which any

Wires were connected and to make a charge accordingly. Bills could be sent in periodically However small the rate of charge might be the revenue would probably be something enormous.

In conclusion I would say that it seems to be that the telephone should immediately be brought prominently before the public as a means of communication between Bankers Merchants Manufacturers Wholesale & retail dealers Dock Companies Gas Companies Water Companies Public Offices Fire Stations Newspaper Offices Hospitals and Public Buildings and for use in Railway Offices in Mines and in Diving operations

Arrangements should also be speedily concluded for the use of the Telephone in the Army and Navy and by the postal Telegraph Department.

Although there is a great field for the Telephone in the immediate present I believe there is a still greater in the future.

By bearing in mind the great objects to be ultimately achieved I believe that this telephone Company can not only secure for itself a business of the most remunerative kind but also benefit the public in a way that has never previously been attempted.

I am Gentlemen
Your obedient Servant
Alexander Graham Bell

WASHINGTON, D. C.,
October 26, 1885.

To the Hon. A. H. GARLAND,
 Attorney-General of the United States.

SIR: I live in Washington, within a few minutes walk of your office, and have resided here for several years. I think you know this.

An instrument which I have furnished to science and the arts has made my name widely known.

I have asserted, and again assert, that I am the first inventor of it. I have received honors which would not have been bestowed had not that assertion been believed; and when my rights have been invaded, the courts of the United States have affirmed the truth of my claim and the validity of my patent.

Three weeks ago I returned from a long journey and found every newspaper proclaiming to the world, on the authority of your Department, that my assertion that I am that first inventor is false, and has always been known by me to be false; and that the patent which the United States granted to me for the invention was obtained by false and fraudulent suggestion and concealment of my part. I find myself charged by your Department with perjury and fraud.

This charge, the newspapers inform me, was made in a suit brought, not in Washington, where you know I reside, nor in Massachusetts, where the Company which is the sole owner of my patent has its domicile—but in Tennessee, where I never have been, and where the owner of my patent has not even an office or an officer.

The suit was brought at the instigation of convicted infringers of my patent; the counsel appointed to prosecute it are their counsel; and its pendency was announced by a United States District Attorney at Baltimore, as a reason why a United States court there should not hear, against one of those infringers, a motion for injunction which it had previously ordered to be heard.

The infringer, which so attempts to use this action of your Department, to shield itself from judgment, is a corporation of which you were one of the organizers, one of the largest stockholders, director, and counsel.

I seek for some explanation. I find that the explanation you gave was so far from sufficient, that upon it alone the suit was instantly dropped. But the very letter of the Solicitor-General, which announced that the suit was withdrawn because the charges were made without proper examination, re-asserted them.

I deny every one of these charges. The official records absolutely disprove them. I propose to call your attention to those records.

The bill, authorized by your Department, alleges (p. 21) that "Prof. Phillip Reis, a German scientist," "was in fact the true and original inventor and discoverer of the electric transmission of sound, speech, and conversation, and the method thereof;" that (p. 17) "at least as early as 1862," by "instruments, apparatus, and methods," invented by said Reis, he *"practised the art of electric transmission, both of articulate speech and musical tones;"* that this art "had long prior to said Bell's pretended invention been openly and publicly practised, not only by the said Reis, but by others," by the "method, means, and instruments so invented and described by said Reis, or by the use of apparatus * * * described or sold to them by said Reis," and that this was accomplished by the "same form of currents, and in the same manner and use of application of electricity," shown in my patent.

Your District-Attorney's bill alleges (p. 18) that said art

"had been since the year 1862 not only publicly known and practised by others than him, said Bell, but the apparatus necessary therefor, and the method or principle, as well as practical direction for the electric transmission of both tones, musical notes, and speech, had been published and described in such minute details in a very large number of printed publications published both in the United States and in foreign countries, that from said publications any person reasonably skilled in the art of applied electricity could of himself construct the instruments, apparatus, or devices necessary for the electric transmission of sound and speech, and so apply electric currents thereto and therewith, as to operate said prior invention and art, and transmit electrically both tones and articulate speech."

It enumerates seventy-five publications and patents, each of which, so it alleges, contains this information.

The bill adds (p. 16) that everything shown or claimed in my patent was "in truth and in fact old and well-known to all scientists and physicists conversant with science so far as it relates to electricity."

The bill says (p. 15) that all this was so well-known as the state of the art that I "cannot be heard to say" that I was ignorant of it.

Whereupon your District-Attorney charges me with perjury and fraud, because I swore that what he calls so well-known an invention was my own, and because, so he avers, I

"Deliberately, deceitfully, and knowingly contrived, and until after said letters-patent were actually allowed and delivered to him, did actually conceal from the said examiner, and from the officers of the United States in the said Patent Office, charged with the examination of his said pretended discovery and invention, all

272

knowledge and information of the true state of the art in that behalf and respect."

I am charged with "concealing from the Patent Office" the existence of an art alleged to have been at that time well known to the whole civilized world, and long openly and publicly practised.

You, sir, would not have made that assertion, because you personally know that before my patent no such art existed.

Only last year your associates in the Pan-Electric enterprise, in the name of the *Washington Telephone Company,* by a printed circular, headed with the names of General Bradley T. Johnson and Mr. Casey Young, publicly asserted the following:

> "The surprising rapidity with which the telephone has been adopted as a means of communicating intelligence between distant points surpasses all the wonders of this wonderful age. *It was first exhibited to the public at the Centennial Exposition in 1876, and the first line for practical use was constructed in Boston in 1877.* To-day, after a lapse of but seven years, there are more than 250,000 instruments in use and the number is increasing at the rate of over 50,000 a year."

That "first" exhibition was *my* exhibition; that "first" line was *my* line. Every one of those instruments was made under my patent, stamped with its date, and furnished by its owners.

Such was the origin of the art.

When I filed my application I asserted that I was the first inventor of that art. I believed so then; I know so now.

In 1876 the United States appointed, to examine the electrical inventions at the Centennial Exposition, an international board of judges, who certainly were "scientists and physicists conversant with science in so far as the same relates to electricity." At their head was Joseph Henry. I exhibited my telephone to them on the 25th of June, 1876, and they heard speech electrically produced. They took my apparatus to their consultation room, and used it by themselves. In their official report, drawn by Prof. Henry, after referring to the transmission of musical notes by current interrupters, they said (Dowd case, vol. ii, p. 32:)

> "The telephone of Mr. Bell aims at a still more remarkable result—that of transmitting audible speech through long telegraphic lines." * * * To understand this wonderful result," &c., &c., "This telephone was exhibited in operation at the Centennial Exhibition, and was considered by the judges the greatest marvel hitherto achieved by the electric telegraph. The invention is yet in its infancy, and is susceptible of great improvements."

273

The most important of the Reis publications were in the Smithsonian library. Prof. Henry had for two years had a Reis apparatus in his cabinets there. Yet to him and his associates a speaking telephone of feeble sort, constituted in June, 1876, *"the greatest marvel hitherto achieved by the electric telegraph."*

Yet your District-Attorney asserts that speaking telephones had long been so well-known that I could not be heard to say that I was ignorant of them.

For not telling to the Patent Office what Professor Henry and his associate judges at the Centennial did not know, I am accused by your Department of "cunning concealment" and wilful perjury.

The Patent Office and the courts, in many contested cases, after fully considering every Reis publication mentioned in your District-Attorney's bill, have decided that the art of transmitting speech by electricity *was not there to be found,* and did not exist before the date of my invention. (Telephone Interferences, 30 O. G., 1091; Am. Bell Tel. Co. *v.* Spencer, 8 Fed. Rep., 509; Same *v.* Dolbear, 15 Fed. Rep., 448; Same *v.* People's Tel. Co., 22 Fed. Rep., 109; Same *v.* Molecular Tel. Co., Fed. Rep.,—)

Yet I am charged with fraud and deceit for not telling the Patent Office in 1876, that which the Patent Office itself and all the courts, after careful and elaborate investigation, have since decided to be untrue.

How could the United States make that charge against me? How can the Washington Telephone Co.—the author of that circular—now ask you to repeat that charge?

In each and every case, the Patent Office and the courts have decided that I am the first inventor of the electric speaking telephone.

After seven days of hearing, the Commissioner of Patents decided:

"He (Bell) invented the art or method, and also an apparatus by which said art or method could be practised. His claim as an inventor is for having done that thing in its entirety." ° ° ° At the Centennial Exhibition "the world recognized Bell as the first inventor of a speaking telephone. The indications are that it was not until the promised reward for so important a public service became visible that his claim to priority was called in question by any of the parties to this interference." (30 Official Gazette, 1091.)

Your District-Attorney's bill charges another fraud, namely that I, not being the inventor of the speaking telephone, unlawfully saw a caveat of Mr. Elisha Gray which described one, and afterwards by an amendment to my application then pending, wrote this into my specification, and obtained the patent for what I had not invented, but had so stolen.

Your District-Attorney alleges (p. 8) that the Examiner in the Patent

Office, on Feb. 19, 1876, "wrongfully gave to the said Bell information of the matter contained and set forth in the said caveat of Gray," and (p. 9) that the said Bell, "being well aware of the nature of the invention of the said Gray, as disclosed in said caveat, through and by reason of the betrayal of confidence by the said Examiner," thereupon, for the purpose of stealing that which, so the bill avers, Gray had invented and Bell had not, did, on February 29, make a certain amendment to his specification, and (p. 10) that the Examiner allowed this, "well knowing that the said amendment was suggested by the unlawful disclosures made by him touching the matters contained and set forth in the said caveat of the said Gray."

I specifically deny that I was indebted to Mr. Gray for any part of the invention described in my patent.

The original files of the Patent Office show that the description of the speaking telephone stands in my patent today *exactly as it stood in my application written and filed before Mr. Gray's caveat existed.*

The records in the Patent Office show the following facts:

1. On January 20, 1876, I swore to my application in Boston, where I then resided.

2. On the morning of February 14, Messrs. Pollok & Bailey, my solicitors in Washington, filed it in the Patent Office.

3. On the afternoon of that day Mr. Gray swore to and filed a caveat, suggesting the idea of a telephone, and swearing that he was "engaged in making experiments for the purpose of perfecting the same, preparatory to applying for letters-patent therefor."

4. On February 19 the Office gave the notices customary in cases where there is a prior caveat on file—one to Mr. Gray, care of his solicitor in Washington, informing him that an application had been filed which apparently interfered with his caveat, and one to Messrs. Pollok & Bailey, informing us that "the 1st, 4th, and 5th clauses of claim relate to matters described in a pending caveat," and that my application was therefor "suspended for ninety days, as required by law."

All this was according to law and the established practice of the Office.

I am advised that R. S. 4903 and the rules of the Patent Office required the notice to inform me what claims were interfered with, in order that I might drop them if I saw fit, and the affidavit of the examiner, Wilbur, filed with you by my opponents, states that such was the practice of the Office.

I am advised that according to R. S. 4902 a caveat is operative "from the filing thereof." Upon a written petition filed by my solicitors the Commissioner ascertained that the caveat was filed some hours later than my application, and, by a written opinion and order, decided that the notice of interference should be withdrawn. This was done February 25, 1876, and the patent issued March 7, 1876.

But the notice of interference *had* been sent to Mr. Gray on February 19. He has since testified that he received it. (Dowd Case, vol. 1, p. 136.) It is a fact that he had ample opportunity to file an application after the withdrawal of the notice, and thereby contest my claims, had he been so minded; but he and his counsel deliberately determined not to do so.

I was not in Washington, and had no personal knowledge of these proceedings until after the notice had been withdrawn; but I am advised and believe that all of them were exactly according to law. Whether they were so or were not, it was not a fraud on my part nor on the part of my solicitors for them to receive and read the official communication sent to them by the Commissioner, nor for them to ask his personal and careful examination into the exact truth of the matter, and to abide by his formal judgment placed on the public records of the Office.

The gravamen of the charge in the bill lies in the suggestion that in consequence of unlawful information (alleged, but untruthfully, to have been given by the Examiner to me) *I changed my application so as to include the invention of Gray, and thus stole from him the speaking telephone.*

The original official records of the Patent Office absolutely disprove the possibility of the act alleged.

1. No such change was made. All the part of my specification which describes or claims the speaking telephone stands in my patent exactly as it was written in the original application now in the files of the Office, and placed there on the morning of February 14, 1876, *before Mr. Gray's caveat existed.*

2. That description and claim as filed, had been explicitly stated by the Office to set forth the invention described in Mr. Gray's caveat. It was on that ground that the interference was declared, and it could not have been declared on any other.

The only change ever made in my specification was the following: On February 29, upon an oral suggestion of the Examiner, which he made to me personally, and in order to give greater precision to the language, I added an explanation of the difference which existed between the electrical changes described in an *application of my own of a year earlier*, and those *already* described in my specification in question. This amendment was drafted by me personally. Of course, I did not reswear to the words, as they added nothing, and the Office did not ask me to do so.

The only amendment ever made consisted in inserting in the specification the words following that are printed in Roman letters; the part in italics was in my application as originally filed:

"In a pending application for letters-patent, filed in the U. S. Patent Office February 25, 1875, I have described two ways of producing the intermittent current, the one by actual make and break of contact, the other by alternately increasing and diminishing the intensity of the current, without actually breaking the circuit; the current produced by the latter method I shall term, for distinction's sake, a pulsatory current.

"*My present invention consists in the employment of a vibratory* or undulatory *current of electricity, in contradistinction to a merely* intermittent or pulsatory *current, and of a method of and apparatus for producing electrical undulations upon the line wire.*

"The distinction between an undulatory and a pulsatory current will be understood by considering that electrical pulsations are caused by sudden or instantaneous changes of intensity, and that electrical undulations result from gradual changes of intensity exactly analogous to the changes in the density of air occasioned by simple pendulous vibrations. The electrical movement, like the aerial motion, can be represented by a sinusoidal curve, or by the resultant of several sinusoidal curves.

"Intermittent or pulsatory and undulatory currents may be of two kinds, accordingly as the successive impulses have all the same polarity, or are alternately positive and negative."

"Page 13, line 15, after '*synonymously,*' insert 'and in contradistinction to the terms 'intermittent' and 'pulsatory.'"

"In the 4th claim erase the word '*alternately,*' whenever it occurs, and substitute therefor the word 'gradually.'"

But the speaking telephone stands described and claimed in my patent in the exact words of my original application, no one of which has ever been changed. After describing the apparatus fig. 5, which is a multiple telegraph, and nothing else, it describes the instrument fig. 7, which is a speaking telephone, and nothing else, as follows:

Fig. 7.

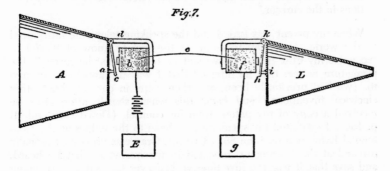

"The armature c, Fig. 7, is fastened loosely by one extremity to the uncovered leg, d, of the electro-magnet, b, and its other extremity is attached to the centre of a stretched membrane, a. A cone, A, is used to converge sound vibrations upon the membrane. When a sound is uttered in the cone the membrane, a, is set in vibration, the armature, c, is forced to partake of the motion, and thus electrical undulations are created upon the circuit, E, b, e, f, g. These undulations are similar in form to the air vibrations caused by the sound—that is, they are represented graphically by similar curves.

"The undulatory current passing through the electro-magnet, f, influences its armature, h, to copy the motion of the armature c. A similar sound to that uttered into A is then heard to proceed from L."

* * * * * *

Claim 5. "The method of, and apparatus for, transmitting vocal or other sounds telegraphically, as herein described, by causing electrical undulations, similar in form to the vibrations of the air accompanying the said vocal or other sounds, substantially as set forth."

The variable resistance transmitter is also described in my patent in the same words as in my original application, as follows:

"Electrical undulations may also be caused by alternately increasing and diminishing the resistance of the circuit, or by alternately increasing and diminishing the power of the battery." * * * "The external resistance may also be varied. For instance, let mercury or some other liquid form part of a voltaic circuit, then the more deeply the conducting wire is immersed in the mercury or other liquid the less resistance does the liquid offer to the passage of the current. Hence, the vibration of the conducting wire in mercury or other liquid included in the circuit occasions undulations in the current."

When my patent was issued and the speaking telephone introduced to the world, Mr. Gray was one of the first to know of it and to recognize my title. In the Dowd suit he testified that he received the premature notice, that he inferred that I was the applicant referred to, (we had previously been, and then were, in conflict about other electrical inventions,) and *knew* this when, shortly afterwards, he received a copy of my patent from his counsel. (Dowd case, vol. 1, p. 136.) I exhibited my speaking telephone to the judges at the Centennial Exhibition on June 25, 1876; Mr. Gray was present, personally listened at the instrument, repeated to the audience what he heard, and says that it was the first time he had ever listened at a speaking telephone. (Dowd case, vol. 1, p. 138.) About the same time he received (so he testifies) my lecture of May 10, 1876, in which I again described various forms of it, including some liquid transmitters. (Dowd case, vol. 1, p. 125.)

Mr. Gray received a copy of my second patent (186,787) January 30, 1877, as soon as it was issued. He wrote me on February 21, 1877, that he was about to give a lecture on his musical telegraph; that he had a copy of my last patent, and asked my permission to construct and exhibit *my* apparatus, adding, (Dowd case, vol. 1, p. 146:)

"I should explain it as *your* method, and not mine, although the Office records show a description of the talking telegraph filed by me the same day yours was filed. *The description is substantially the same as yours.* I was unfortunate in being an hour or two behind you. There is no evidence that either knew that the other was working in this direction."

Mr. Gray, therefore, then recognized that the *"talking telegraph"* was as well described in my specification written in January as in his caveat written in February, and filed "an hour or two" after my application, and that all that was in my two patents was justly mine and not his.

March 5, 1877, he wrote me again, (*ib.*, 148):

"I gave you full credit for the talking feature of the telephone, as you may have seen in the associated press dispatch that was sent to all the papers in the country, in my lecture in McCormick Hall, February 27, 1877. * * * * I described your apparatus at length by diagram."

Mr. Gray exhibited his musical telegraph (which at that time he termed a "telephone") at a public meeting in New York in April, 1877. The *New York Tribune* of April 3, 1877, contained the following, (Dowd case, vol. 1, p. 156:)

"After the first part of the programme had been executed, Mr. Elisha Gray came forward and addressed the audience. He was aware that great confusion existed in the popular mind as to what this telephone could perform; in particular, it had been confounded with the speaking telephone invented by Prof. A. Graham Bell, of Boston. Prof. Bell, Mr. Gray said, was present in the audience."

I was present; I heard Mr. Gray make that statement; and in the Dowd case it was admitted that he did make it.

My second patent of January, 1877, was for improvements in magneto instruments. Mr. Gray testified in the Dowd case (vol. 1, p. 127), that he believed the magneto-transmitter to be entirely new with me.

All the facts relating to this matter have been more than once judicially investigated by the courts and by the Patent Office, and the decision has invariably been in my favor. In 1877, when the Western Union Telegraph Company was sued for infringing my patent, it bought Gray's claims, filed an application for a patent on his caveat, and an interference was declared between that application and others of his applications and my two patents. After long hearings, the Office decided that he had no right to claim the speaking telephone, and that I was the first inventor of it. It also decided particularly that the liquid transmitter and the receiver of Mr. Gray's caveat were described *in my previously-drawn application, and that I was the first inventor of them.* (15 O. G., 776; 30 O. G., 1091.)

Your District Attorney's bill alleges that my patent neither describes nor claims a speaking telephone.

I need not discuss that question. My patent is now before the Supreme Court, in Dolbear's case (No. 368 on the docket) and in the Molecular case; and with it are all the Reis publications named in the bill. You need not bring a new suit in Memphis to try that; but the question has been often decided. Indeed, your bill so shows.

The said letters-patent were granted and issued to me on the 7th day of March, 1876, and were numbered 174,465. Now, the bill of complaint itself confesses that on the 19th day of the preceding month my application had been declared to interfere with the caveat of Mr. Elisha Gray, referred to above. The records of the Patent Office show that this caveat advanced but one single claim, namely, a claim to the invention of "the art of transmitting *vocal sounds or conversations telegraphically* through an electric circuit."

It thus appears from the records and from the bill of complaint itself that the Patent Office knew on and before the date of my patent that my claims included, and were intended to include, a speaking telephone, and ordered an interference on that ground alone.

Mr. Gray also wrote that he found it in my patent (p. 12, *supra*.)

In the first contest which arose the Commissioner of Patents so decided. Referring to the passages in my specification, which have been quoted above, he said (15 O. G. 776:)

"Nor is the fifth claim of Bell's patent (No. 174,465) limited to the transmission of other sounds than articulate speech; for, if the word 'vocal' in the claim itself could leave the question in doubt, the description in the application would remove such doubt."

The meaning and scope of my patent has been exhaustively determined in that opinion and in the decision of Mr. Justice Gray in Dolbear's case (15 Fed. R., 509).

My patent is nine years and a half old. I have referred to the honors I have received, and to the decisions by which my title has been sustained. These judicial investigations have been aided by more than thirty printed volumes of evidence and arguments; more than fifty days of oral argument; while more than thirty counsel and twenty-five experts have given the aid of their skill to my opponents. Every contest has resulted in a decision in my favor.

How came the United States to prefer against me charges which have so often been refuted by its courts, and which its own official records in its Patent Office absolutely disprove?

When my patent was seven years old, and after it had been sustained by the courts, and when 250,000 of my instruments were in use, a group of Senators and Representatives of Congress formed the "Pan-Electric Telephone Company," which proposed to "monetize" a paper capital of five million dollars by using telephones like those which had been adjudged to infringe my patent. (See their contracts recorded in the Patent Office May 21, 1884, and Sept. 17, 1884.)

The Philadelphia branch which this company had agreed to defend was perpetually enjoined June 26, 1885. At the time the application was made to you, a suit for infringing my patent had been brought at Baltimore against the *Pan-Electric Co.*, its branch, the *Washington Telephone Co.*, Gen. Bradley T Johnson, as president of the latter, and Mr. Casey Young as director of both. In that suit a motion for an injunction was ordered to be heard on Sept. 15, 1885.

About 1883-4 the *National Improved Telephone Co.* was formed by parties interested in the Pan-Electric Co. for similar purposes with a similar capital. Its Pittsburg branch, defended by it, was enjoined *pendente lite* July 8, 1885. That suit is to-day ready for final hearing.

The grounds of attack presented by your District-Attorney's bill could be heard in those suits,—and speedily, if a trial of them was what was wanted. But the infringers preferred, if they could, to escape or postpone a hearing.

Your letter to the President states how they went about it:

"Some day last summer, I do not recollect the precise day, some gentlemen approached me stating that they desired to *make application in the name of this company, of which I was a stockholder and attorney*, FOR THE USE OF THE NAME OF THE UNITED STATES to test the validity of the Bell telephone patents in the courts. Those gentlemen were Mr. Casey Young, Col. George W. Gantt, Mr. Van Benthuysen, and one other gentleman whose name I do not recollect."

Mr Casey Young was director, secretary and counsel of the *Pan-Electric* and director of the *Washington Telephone Co.*, and personally a defendant in the Baltimore suit; Mr. Gantt had already appeared as counsel for the *Pan-Electric* in a suit between it and the owners of my patent.

You took no official action, but left them with the understanding that the application should be renewed, for in your letter to the President you say:

> "After the interview with the gentlemen already spoken of, I supposed they would come to me with an application after the statement I had made, to either refer the matter to the Solicitor-General or to present it to you as the head of the executive department of the Government for your consideration."

So they came again to your Department to borrow from it the name of the United States.

On August 27 you left Washington for a vacation.

Saturday, Aug. 29, is the date of an application to the District-Attorney at Memphis, (the domicile of the Pan-Electric Co.,) asking him to bring a suit there, praying that my patent be cancelled. On Monday, Aug. 31, his mind was made up, and he sent a letter to your Department asking permission.

This letter reached your Department on Sept. 2, and was received by the Acting Attorney-General, (the Hon. John Goode.) At the same time the Acting Attorney-General was waited upon by Senator Isham G. Harris and Congressman Casey Young, vice-president and secretary and counsel for your Company, "The Pan-Electric."

On the next morning they called again, and were informed that the Acting Attorney-General had already sent the permission asked for— so swift was justice! He had also ordered that special counsel assist in my prosecution, and had named for that purpose Messrs. Young, Gantt, and Wright, counsel for the Pan-Electric Co., and Mr. Beckwith, counsel for the National Improved Telephone Co.—on condition that the United States be at no expense for their fees—so sure was he that the United States had no interest which would justify an appropriation.

In the ordinary course of mail this order would reach Memphis on September 5. On September 9 the bill and schedules of forty printed pages were filed in print, and the imprint on them is that of a New Orleans printer.

Do not these dates and this unwonted swiftness show that the private prosecutors had prepared the papers beforehand, and looked upon the application to your Department as a mere form, the out come of which they felt sure of; and that no real examination could have been made or was expected?

But why this haste?

On the same day that the bill was filed a special dispatch from

Memphis was sent to the Chicago *Inter-Ocean,* and published on Sept. 10, which said:

> "The attorneys here claim that the effect of filing the bill will be to suspend proceedings of the Western Union and Bell Telephone people against other telephone companies."

Whereupon those who procured the bill to be filed proceeded to use it in a suit in which the Pan-Electric Co. and its associates, and one of those very attorneys, were defendants. On Sept. 15, when the motion of the Bell Company for an injunction against the Pan-Electric Company and the Washington Telephone Company came on to be heard at Baltimore, Mr. Sterling (who is the United States District-Attorney) said to the Court:

> "Now it does strike me that so far as a preliminary injunction is concerned, it is a valid objection to a preliminary injunction if, since these other decisions, the United States has filed a bill alleging the patent to be void, alleging it to have been fraudulently obtained, and asking the court to declare it void. * * * It does seem to me to attack the validity of this patent in a form and in a mode of action which is an absolute bar to the proper exercise of legal discretion in granting a preliminary injunction.

To enable the promoters to say that the action was made in your absence, and at the same time have it in season to use at Baltimore, there was no time even for the usual reference to the Interior Department—and none was made.

I am informed that only four such suits had been brought in the history of this country, and, in one of them, *United States* v. *Fraser,* (22 Fed. R., 106), a suit to cancel a patent for invention, the court said:

> "It is now only at the instance of parties who are specially and directly interested in their defeat, and who, by the showing of the bill, have a complete defence against both these patents, that the name of the Government is lent to these contestants to attack these patents, and that only upon the guarantee that the Governmen is to incur no costs. * * * Would it not be better to leave the attack upon such patents as have been obtained by false suggestions where they have heretofore been left, as defences to the validity of the patents."

The Court there decided that a court of equity would not entertain a bill brought under such circumstances, and dismissed the suit.

United States v. *Throckmorton,* (98 U. S., 71,) was a suit to cancel a land patent. The Supreme Court said:

283

"It would be a very dangerous doctrine and one threatening the title to millions of acres of land held by patent from the Government, if any man who had a grudge or claim against his neighbor can, by indemnifying the Government for costs and furnishing the needed stimulus to the District-Attorney, institute a suit in chancery in the name of the United States to declare the patent void."

Whereupon, the Court decided that it would not entertain such a suit, and dismissed it on demurrer.

I am not blind to the real character of the proceeding sanctioned by your Department. It was not your personal act. It did not even originate with your Department. It was due to convicted infringers. There is reason to fear that they hoped the influence of your known personal interest would make action more speedy and scrutiny less careful. Plainly a proper examination would have defeated their scheme.

I know that the official order was given by the Acting Attorney-General (Solicitor-General) while you were absent from Washington. But the facts remain that you knew of the first application and expected its renewal; that the action was unusual, taken with extraordinary haste, without time for suitable examination, without the usual reference, in the face of repeated decisions of the courts, in spite of absolute disproof from the records; that it selected a forum unlawful and oppressive; that it was in substance at the instance (in part at least) of a corporation in which the head of the Department had an enormous interest, urged by the personal presence of your co-directors, and instantly (though unavailingly) employed in the attempt to protect that corporation from a trial where each allegation, *if true*, would be a defence for it.

I have no fear that my recognized position as the inventor of the speaking telephone will be impaired in the estimation of scientific men and the world at large by the attack that has thus been made upon me; but I have not the philosophy to endure with patience the accusation of fraud and perjury brought against me in the name of the Department of Justice of the United States, even though the Patent Office which granted my patent, and the courts which have sustained it, are included in the accusation.

I have, therefore, made this statement of facts to be filed in the records of your office.

I am, sir, yours respectfully,

ALEXANDER GRAHAM BELL.

BIBLIOGRAPHY

DeLand, Fred, *Dumb No Longer*. Volta Bureau; 1908.

Fairchild, David, *The World Was My Garden*. Scribner; 1938.

Green, H. Gordon, *The Silver Dart*. Brunswick Press, Ltd., Fredericton, N. B.; 1959.

Mackenzie, Catherine, *Alexander Graham Bell*. Houghton, Mifflin; 1928.

The Deposition of Alexander Graham Bell in the Suit Brought by the United States to Annul the Bell Patents. American Bell Telephone Company, Boston; 1908.

Watson, Thomas A., *Exploring Life*. D. Appleton; 1926.

ABOUT THE AUTHOR

To write this book Helen Waite spent many weeks meticulously examining the monumental collection of the Hubbard-Bell letters and diaries and other family material, including photographs which Dr. Gilbert Grosvenor had assembled and correlated in the Bell Room of the National Geographic Society in Washington, D. C. Her task also took her to Florida, and—most thrilling of all—to *Beinn Bhreagh,* the home Mabel and Alexander Graham Bell made for themselves in Baddeck, Nova Scotia. She spent eight wonderful days at the Alexander Graham Bell Museum there, she saw the places where Dr. Bell had worked, she sailed the Bras d'Or, talked to the people who had been with the Bells, and finally climbed to the summit of their Beautiful Mountain where they lie buried.

She feels that she grew very well acquainted with the Bells, and hopes that she has captured half of this feeling in the book.